# Having Healthy Honeybees
## *The beekeeping & science*

John McMullan

Northern Bee Books

*To Aoife*

# Having Healthy Honeybees

*The beekeeping & science*

John McMullan

**Northern Bee Books**

Published 2021 by
Northern Bee Books, Scout Bottom Farm,
Mytholmroyd, West Yorkshire HX7 5JS
Tel: 01422 882751 Fax: 01422 886157
www.northernbeebooks.co.uk

ISBN 978-1-912271-90-0





Design and artwork DM Design and Print





**PHOTOGRAPH ACKNOWLEDGEMENTS**
Photographs, copyright of author except for figures, 4.7a (D. Anderson); 4.4a
(Canada DOA); 4.4b (I. Fries); 3.5a, 5.5a, 5.8a (V. Groult); 2.3b (P Gurrie); 6.1a,
b, c, g (IBRA); 2.1a (Institut fur Bienenkunde, Oberursel); 4.13f  (iStockphoto);
4.2d, 4.3b, 4.6a, 4.8a, 4.10a, 4.14a (National Bee Unit, Fera); 3.2a (T Seeley);
2.2a, 5.3e (R. Strba); 4.9a (University Newcastle); 4.1c,6.1b (R. Williams).

# CONTENTS

# Preface

There has been a change in beekeeping since the first edition in 2012. The preoccupation then was the *Varroa* mite, integrated pest management and the book was subtitled *an integrated approach*. Today there is heightened concern about global warming, the natural environment, biodiversity and a need to be guided more by science. The UN has called for 2021 to be the year to 'end the war on nature'. Among beekeepers there is a greater awareness of these issues with a shift towards sustainable beekeeping, non-treatment of colonies, reduced use of chemicals, and concern for pollinators and forage. These concerns are reflected in this second edition. This increased engagement of beekeepers and the questioning of otherwise accepted practices has prompted the title of this edition, *Having Healthy Honeybees – the beekeeping & science.* There are over 220 references listed, many of which are on open access.

I would like to thank all those who helped in the preparation of the first edition, in particular the Federation of Irish Beekeepers' Associations who gave great encouragement at each step along the way. This edition builds on the first, and to acknowledge this the preface to the first edition is included here. I would like to thank Professor Gard Otis, Guelph University, for his critical inputs over the past 20 years and Dr John McKillen, AFBI, Northern Ireland for his help in virus testing. Also, members of Fingal North BKA for their support over many years. Thanks also to Jeremy Burbidge, Northern Bee Books, for publishing this edition, and also David Miller for typesetting and design. And finally, Rosaleen for putting up with my beekeeping ways over the years.

John McMullan Ph.D.
May 2021

**Preface** (to first edition)

The health of honeybees is under threat today from more diseases than at any time in the history of beekeeping. This has given rise to a lot of information on diseases, much of it very detailed, some of it not well founded and seldom available from a single source. This book uses an integrated approach; identifying and bringing together the factors that have a bearing on bee health, and also combining the science with practical beekeeping. The aim is to help beekeepers to establish healthy colonies, assess their condition and take appropriate action. It attempts to provide beekeepers with a concise format and in particular help the large number of new, small-scale beekeepers who have taken up the craft. As beekeeping is an old craft with a long tradition, the text retains the old designation of "honeybee" and not the usual scientific one of "honey bee".

I would like to thank the Federation of Irish Beekeepers' Associations for publishing this book, and in particular the Secretary, Mr Michael Gleeson, for his help and encouragement at each stage along the way. My gratitude is also due to those who gave willingly of their time to read all or part of the manuscript and provided valuable feedback; Mr Norman Walsh, MBE, Ms. Megan Seymour, Professor Robert Paxton, and Dr. Mary Coffey. I am indebted to Dr Eoghan Mac Giolla Coda who at short notice reviewed and proofread the manuscript. Ms. Lucy Proctor undertook the graphics and typesetting, with great patience and professionalism.

Trinity College Dublin has been a great source of help and inspiration over the years and for that I am deeply grateful. Having open access to colonies of members of Fingal North Dublin BKA over a long period has greatly assisted my beekeeping and research. In particular, thanks is due to John and Dorothy Stapleton, who for a considerable time have effectively given me "the run" of their farm for my beekeeping activities.

Lastly, as beekeepers we sometimes take for granted the help given by our families to enable us to fully engage in our craft, and for this I thank my wife Rosaleen for her support and encouragement over many years.

John McMullan, Ph.D.
January 2012

# PART 1 | Introduction

There is currently a lot of literature available to beekeepers, increasingly through the Internet, some of it contradictory but much of it offering the beekeeper many alternative views and options. This lack of clarity can result in the neglect of the honeybee, often at the hands of well-meaning beekeepers. As beekeepers, our primary aim must be to fulfill our *duty of care* to the honeybee. This book therefore has an emphasis on keeping it simple to assist beekeepers to get a grasp of the essentials and be in a position to take action.

The book acknowledges the special needs of small-scale beekeepers, who despite making up the vast bulk of the beekeeping community, are often a neglected beekeeping constituency. In most European countries, over 90% of beekeepers have fewer than ten colonies. Their small scale and limited colony-years of experience, and in many cases with a sense of isolation, makes them more dependent on beekeeping literature for information. The ways of larger scale beekeepers with an emphasis on honey output and speed should not dictate. Recent evidence would suggest that large scale (professional) beekeepers have higher disease levels than smaller outfits[37].

Despite the recent gloom that has surrounded beekeeping, largely following the high level of colony mortality, beekeeping still retains its old-craft charm. There has been a large increase in new entrants to beekeeping, and a new vigour is apparent at beekeeping gatherings. For most of us starting out in beekeeping, the honeybee is a strange animal, and we must try to understand its nature and ways if we are to have a happy relationship. Beekeepers, and particularly beginner-beekeepers, should realize that they could be exposed to practices that are *ad hoc* or hearsay and are not evidence based.

Accordingly, there has been an emphasis throughout the text in providing the reader with the reason, the science involved, and extensive references are provided most of which have been peer-reviewed. The hope is that this will move the beekeeper towards greater knowledge with understanding, which is a much more durable commodity than knowledge alone and should help to prevent a drift in practices over time. While this is particularly important for the beginner-beekeeper, as beekeepers we can all get set in our ways and continue to use bad practices or harbour misconceptions even where there is clear knowledge to the contrary. We should be aware of the wisdom in the comment of the Nobel laureate Max Planck, which can equally apply to beekeepers: *A new scientific truth does not triumph by convincing its opponents and making them see the light, but rather because its opponents eventually die, and a new generation grows up that is familiar with it*[163]. A case in point was the 'confusion' between Chronic Bee Paralysis Virus and tracheal mites *Acarapis woodi,* in spite of clear scientific evidence (para. 4.9).

On the other hand, science could have a greater focus on serving the beekeeping community. A review of research relating to honeybees and beekeeping for the ten-year period up to 2013 showed that the vast bulk of it dealt with basic or pure research rather than applied[142]. Most *modern* beekeeping practices have been around for a long time, even shook swarms have been with us for over 100 years[59]. It would help if beekeepers were more involved in critically appraising research findings and have a stronger voice in shaping the direction of new research. The increased relevance would also benefit the scientific community. In this regard, greater use could be made of the international colony loss reporting system COLOSS (Prevention of Honey Bee COlony LOSSes).

The author has felt for some time that honeybee health and diseases are generally sparsely covered in beekeeping texts, probably as diseases can be an off-putting subject. However, their importance is such that bee health is influenced by almost all beekeeping actions. Bee health is given particular emphasis here to remind beekeepers that disease is not some passive affliction that *happens* to bees from time to time. Rather, diseases are caused by living organisms (parasites/pathogens) that undertake their lifecycles in or on bees while at the same time being nourished by them. This visualization of a dynamic situation should help beekeepers to better understand and identify the conditions in the field.

A brief history of beekeeping and diseases is given in Part 2. Some aspects of bee biology are included, enough to provide a context, and the honeybee as a pollinator is also covered. The glossary at the end should also help to clarify terms.

Part 3 deals with the critical area of the beekeeping set-up. If beekeepers do not get this right in the first instance, they will have an uphill struggle from the outset.

An introductory note on diseases is provided in Part 4 to explain terms and to provide background to the later discussion. This part deals with each of the major diseases currently affecting honeybees. They are grouped into the four main parasite categories: bacteria, fungi (including microsporidia), mites (including beetles) and viruses. To improve clarity, each disease is treated under a set of four headings: cause, signs/diagnosis, virulence/spread and treatment. To help the beekeeper to recognize the various conditions, illustrations are used extensively throughout the text. The author has not given his colonies treatments of any kind for over ten years. For those who hopefully are phasing out chemical treatments, some options are given for the short term. It should be noted that everything is subject to the *National Policy* in each country, including regulations pertaining to the use of approved chemicals.

There is an increasing need for beekeepers to be able to establish cause of death particularly during the winter period. This has been heightened by the recent spate of colony deaths and the requests for beekeepers worldwide to make returns through national and international systems, such as COLOSS. This book can therefore act as reference material to help improve consistency in establishing the cause of death.

Part 5 covers the all-important subject of working with the bees. Using the native (indigenous) bees that are in harmony with their environment allows for sustainable beekeeping. Keeping manipulations to a minimum and being gentle in handling should hopefully allow progress on a course that allows for the use of bare-handed manipulation, surely one of the great pleasures in beekeeping.

Forage and honey are covered in Part 6. Without forage there can be no bees and no beekeeping. The availability of good supplies of nectar and pollen is essential for bee health and a good supply of surplus honey. The influence of forage on the taste, colour and aroma of honey always adds interest to our beekeeping. The forage also influences the health and medicinal benefits of our honey[110].

Finally, an extensive glossary is included at the end which defines most of the terms that beekeepers will encounter. The items are referenced to the specific paragraph(s) in the text.

# PART 2 | Background

## 2.1 A brief history of beekeeping & diseases

It is generally accepted that honeybees evolved in Africa or Asia over 50 million years ago[51,210]. They were cavity dwellers and mostly occupied caves and hollows in trees. From early times, mankind has interacted with bees through robbing (honey hunting) colonies in the wild[47].

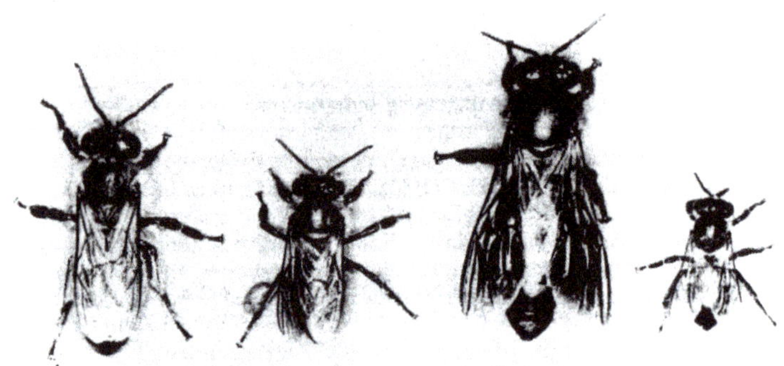

**Figure 2.1a** The four main species in the *Apis* genus are illustrated to show their relative sizes; from left; *Apis mellifera, Apis cerana, Apis dorsata* and *Apis florea.*

Honeybees are classified in the genus *Apis*, and have over ten species, the four main ones **(Fig. 2.1a)** being the common or European honeybee (*Apis mellifera*), the closely related Asian honeybee (*Apis cerana*), the tropical giant honeybee (*Apis dorsata*) and the tropical dwarf honeybee (*Apis florea*)[44,218].

The European honeybee has a number of subspecies, the most important being the North European dark bee, *A. m. mellifera*, and its more southerly counterparts: the Italian, *A. m. ligustica*, the Carniolan, *A. m. carnica* and the Caucasian, *A. m. caucasica*[218].

Figure 2.1b Arch stone dated 1689 from Bremore Castle, adjacent to St. Molagga's church in north County Dublin, that appears to depict a monk on the right holding a skep with bees flying towards it.

Figure 2.1c Bee boles at Bremore Castle today in which skeps of bees were kept in the seventeenth century.

The first form of hive beekeeping in Europe involved log hives, which later gave way to skeps, conically shaped baskets, usually made from braided straw, in which honeybees were kept[47]. In Ireland, hive beekeeping was reputed to have been introduced by a sixth century monk, St. Molagga, who built a church, *Lann Beachaire* (the Church of the Beekeeper) near Balbriggan, in County Dublin[48,212]. An arch stone dated 1689 in the adjacent Bremore Castle depicts a monk holding a bell (or a skep) and bees flying towards it **(Fig. 2.1b)**. Skeps were kept in sheltered locations or bee boles particularly during the winter months **(Fig. 2.1c)**.

Skep beekeeping was crude and required a large proportion of the bee colonies to be killed annually to enable the honey crop to be extracted. However, the killing of weak colonies and the frequent renewal of brood comb played a large part in keeping disease under control, particularly brood diseases. This is reflected in *Instructions for managing bees* published in 1733[196], where diseases are not even mentioned and where the "great enemies to bees" are listed as mice, wasps, moths and earwigs.

The advent of movable frame hives (largely following the work of the Rev Lorenzo Langstroth) from the mid-1800s brought about the replacement of much of the skep beekeeping worldwide in the following fifty years. It made possible a more *modern* approach that allowed the beekeeper to undertake manipulations that were denied with fixed-comb skeps; not least was the ability to examine individual combs for the presence of disease. However, it was not all positive from a disease perspective; the reduction in bee-swarming and brood-comb renewal, along with an increased movement of frames between colonies, contributed to an increase in diseases such as American foul brood (AFB)[88].

In the history of honeybees there have been many reports of bee diseases, some of which claim widespread colony deaths. Aristotle in the fourth century BC described maladies in bees that included hairless black bees, which fits the clinical signs of what we associate today with *Chronic bee paralysis virus*[22]. Some of the earliest reports of bee losses were in Ireland. According to the Annals of Ulster, there was "a mortality of bees" in 950 A.D., and in A.D. 992 "a great mortality upon men (duine-badh), cattle, and bees in Ireland this year" was reported. In the Annals of Ireland, a third bee epidemic was recorded in 1443 A.D.[77].

Disappearing bees were recorded in a number of US states in 1869 and in Australia in 1892. Then just over one hundred years ago (1904) the much-reported colony deaths on the Isle of Wight, just off the south coast of England[1], were first recorded. The bee mortality spread throughout England and Wales, reaching Scotland in 1910 and Ireland in 1912. A tracheal mite, *Acarapis woodi*, was identified in honeybees in Scotland in 1919 and the mite became associated with widespread colony deaths and a condition that has since been referred to as acarine, *acariosis*, tracheal mites as well as "Isle of Wight Disease". The mite was found throughout Europe in the first half of the 20th century and eventually in South and North America (1984) and is currently to be found in all major beekeeping countries with the exception of Australia and New Zealand. It is arguably the most controversial disease in the history of beekeeping[15,1], and its pathology is still disputed.

Bee diseases that we know today have largely been identified since 1900.

They spread progressively throughout the beekeeping world, invariably through the global movement of colonies. AFB, which is one of the most virulent bee diseases, was identified and named in 1906[213] and its sister disease, European foul brood (EFB) in 1912[215], while Chalkbrood was identified in the early 1900s. *Nosema apis* was first isolated and named in 1907[225] and was considered for a short time by the Board of Agriculture in Britain to be the cause of "Isle of Wight Disease". A spate of viruses has been identified starting with *Chronic bee paralysis virus* (CBPV) in 1963[25], signs of which had been described by Aristotle as mentioned above. The number of viruses that has been isolated has been steadily increasing with improved sequencing technologies and by 2020 probably numbers in the order of forty[29,40,173]. Other established viruses include *Sacbrood virus* (SBV), *Acute bee paralysis virus* (ABPV), *Deformed wing virus* (DWV), *Kashmir bee virus* (KBV), *Slow bee paralysis virus* (SBPV) and several other more minor viruses. Some of these, such as DWV, have only become major diseases in conjunction with the macroparasite *Varroa destructor*.

The arrival of *Varroa* mites has had a devastating effect on the European honeybee. The mite had co-existed with the Asian honeybee, *Apis cerana*, but shifted to the European honeybee, *Apis mellifera*, in the first half of the 20th century when *Apis mellifera* was introduced to Asia, and the mite migrated west. This exotic mite and the European honeybee did not have a stable relationship and without treatment most honeybee colonies collapsed within 2-3 years in temperate climates[171]. The wild (feral) colonies were dying, reducing the gene pool and affecting queen mating. This further compounded the mating problems that were being attributed to reduced drone numbers in hive bees and general fertility as a result of chemical treatments. Honeybee viruses were accentuated and other established bee diseases in conjunction with *Varroa* mites were having a profound effect on colonies. Beekeeping had entered a new era.

A new species of the Nosema microsporidian, *Nosema ceranae* Fries[86] has in recent years spread throughout the beekeeping world and appears to add to the established species, *Nosema apis*. This pathogen of the Asian honeybee, *Apis cerana*, shifted host to the European honeybee, *Apis mellifera*. It was recorded in *Apis mellifera* for the first time in Europe in 2005 (in Spain) and was associated with large winter colony losses.

Finally, colony collapse disorder (CCD) has been used to describe a phenomenon of colony losses of devastating proportions that have occurred in the United States since the winter of 2006/2007. Whereas there have generally been increased colony deaths in this period throughout the beekeeping world, they have not conformed to the specific symptoms of CCD, although there may be some common factors. Colonies with CCD in the US suddenly die out with very few bees left in the hive. Throughout history, there have been many terms used to describe disappearing bees; "dwindling", "disappearing disease", "die-off", "melt down", but they have usually been of limited scale and duration. In the case of CCD, there is an emerging view that interactions between honeybee parasites and other stress factors, such as the chemicals used in agriculture/apiculture and poor nutrition, are involved[208]. From time to time there have been claims that CCD exists outside the US, but to date these claims have not been sustained.

## 2.2 The individual honeybee

The honeybee is an arthropod, an animal with a segmented body and jointed limbs. It is classified in the order hymenoptera that also includes wasps, its nearest ancestor[218]. It is thought that its separation from the wasp (that obtained its protein from animal matter) started about 100 million years ago with the first appearance of flowering plants. The evolution of plumose hairs in honeybees along with the corbicula for carrying pollen facilitated the change[145]. The segmentation of the honeybee is in three parts; head, thorax and abdomen (**Fig. 2.2a**).

The head contains the brain, and the systems that facilitate bee-behaviour that include: visual perception by eyes (2 compound, plus 3 simple ocelli lens), smell, taste (2 antennae), mouthpart or proboscis for sucking, mandibles for chewing, and glands mainly for food processing such as the mandibular gland and the hypopharyngeal gland in workers producing lipids, protein and vitamins[191]. The mandibular gland in the queen produces the *queen substance* that regulates hive activity such as swarming. This is one of several gland secretions that produces pheromones.

The central part, the thorax, is the engine of the bee and a large part of its bulk is taken up by the big wing muscles that can propel the bee at an average speed of ~ 25 km per hour with a frequency of over 200 beats per second[218]. Air intake openings (spiracles) and tubes (tracheae) provide air/oxygen to tracheal sacs. The thorax is also the *chassis* of the bee with two pairs of wings and three pairs of legs attached.

Finally, the abdomen contains most of the circulatory (blood/ haemolymph), respiratory (air/oxygen) and alimentary (food) systems. Through these systems the food intake, largely carbohydrates (nectar/honey) and protein (pollen),

**Figure 2.2a** Three castes of the honeybee; queen with white spot, worker below and drone to the right.

**Figure 2.2b** Abdomen showing four wax flakes from the glands on the left side, and two flakes from the four glands on the right.

metabolises. The importance of nutrition will be dealt with in (para. 5.5). The abdomen also contains the rectum/excretory system and a wide range of glands for functions such as: wax production **(Fig.2.2b)**, communication (*Nasanov* scent gland) and sting mechanism in worker. Also, reproduction glands, including the spermatheca in the queen. The rectum in the worker can expand to fill a large part of the abdomen during long cold spells in winter, until the bee can get suitable weather for a cleansing flight.

A cuticle covers the outer surface of the bee and acts as an external skeleton. It is tough, flexible and is a physical protection for the bee. Along with its epicuticle covering it waterproofs the bee, keeping it from being desiccated.

For further details and illustrations on the biology of the honeybee refer to specialist texts[49,53,191,118] and general texts[105,211].

The queen, worker and drone are the three individuals that make up the honeybee colony. Their relative sizes are shown in **Figure 2.2a.** They have unique roles to play in the colony. The queen lays the eggs that will produce the worker bees and drones, and eventually the egg that will replace her after she leaves the colony with a swarm. She mates, over a few days at the beginning of her life, with several drones (10 – 28)[116] and stores the sperm in the spermatheca. Drones develop from eggs that are not fertilized by the queen. During reproduction each of the three individuals pass from egg, to larva and then pupate with different time periods for each. During development queens are fed exclusively with royal jelly while the other two castes are fed initially with brood food, and later with a mixture of brood food plus honey and pollen or stored bee bread. Average development times in days are shown in **Figure 2.2c,** but these times will be influenced by factors such as brood temperature and nutrition[218]. In a laboratory situation the author has shown that worker brood pupated at 30°C (not 34.5°C) will result in an ~5-day delay in bee emergence[130,200].    However, bees raised at this extreme temperature will not be viable. The number of eggs laid per day will vary with time of year and for most subspecies would peak at over 1,000.

The workers are infertile females and are produced from eggs that have been fertilized by the queen. As the name implies, they do most of the tasks in the colony and use a division of labour that is broadly age dependent but with flexibility to respond to actual colony needs. The first three weeks as nurse bees are spent cleaning, feeding young larvae, producing wax, comb building, heating and guarding. After this they will engage in scouting and foraging for nectar and pollen[218]. Foraging is facilitated by the use of scout bees to identify suitable nectar and pollen bearing plants, and the scout's ability to communicate the location of the chosen forage. Using the direction of the sun in the sky, enhanced by its ability to detect polarized and ultraviolet light, the scout can communicate, via a dance, the location of the forage.  The waggle (figure-eight) dance performed on the face of the comb indicates the direction and distance from the hive to the forage. The round dance indicates that the source is close at hand. By dancing on the surface of a swarm, nest-site scout bees also communicate the location of a prospective home for the swarm[218].

|  | Worker | Queen | Drone |
| --- | --- | --- | --- |
| Egg hatched | 3 | 3 | 3 |
| Larva fed and cell capped | 8-9 | 8 | 9-10 |
| Prepupa (5th Moult) | 13 | 11 | 14 |
| Final moult | 20 | 15 | 23 |
| Emerged from cell | 21 | 16 | 24 |
| Ready to mate |  | 20 | 37 |

**Figure 2.2c** Typical average development times in days, after laying of egg, for the three honeybee castes.

The drone's primary role is to mate and fertilise the young queens on their mating flights. Drones are only produced in the hive over a short time-period that is in harmony with the reproductive needs of the colony. They will gather in 'drone congregation areas' to meet virgin queens. While they provide some direct benefits within the hive, for example in the form of heat shielding, they are considered essential to the well being of the colony. Healthy well-adapted drones are critical for colony health in the post *Varroa* era[107]. At the end of the summer, colonies that have their queens (queenright) will evict the drones since at that stage they are a drain on resources.

## 2.3 The colony

Colonies of the European honeybee (*Apis mellifera*) typically increase to over 25.000 individuals during the season (sometimes much more depending on weather, forage and subspecies) and work in harmony creating what can be called a superorganism. In its simplest description it is a large group of individuals working together for a common cause and taking on the appearance of a single entity. More specifically from a biological viewpoint, *It seems correct to classify a group of organisms as a superorganism when the organisms form a cooperative unit to propagate their genes*[182]. The superorganism (honeybee colony) depends on scale for its existence, and in small colonies can come under threat and lose its cohesion. This is particularly critical in temperate climates as demonstrated by the vulnerability of small colonies (nuclei) to stress diseases such as Chalkbrood and Nosema spp. Furthermore, bees that are pupated at reduced temperature have lower cognitive ability, impaired bee dances and lose viability[200]. Accordingly, beekeepers should strive for strong colonies, in particular when going into winter. Due to polyandry, the mating of several drones with the queen, the colony is made up of many patrilines or families of half sisters. These half-sister families can bring diversity to the overall colony and give it advantages such as the ability to respond to dynamic changes in forage as well as diseases[124].

This superorganism performs the many tasks mentioned in para 2.2 above. Not least of these is the ability to perform a precise level of thermoregulation, controlling the temperature and humidity in the hive throughout the year. Cluster temperature measurements that the author took during a cold period in early spring showed that when brood is present in a healthy colony the temperature in the cluster stays within the limits of 34.5±0.5°C. **Figure 2.3a** shows a temperature difference of over 30°C (> 55°F) between mean brood and mean minimum ambient temperature during the period, with a maximum difference in the period of over 40°C (> 70°F). This heating is achieved by the workers using the large 'wing' muscles in their thorax to convert the chemical energy in honey to a heat energy output by shivering rather than a mechanical (kinetic) energy output when in flight. They can also change the size of the cluster to increase or reduce its surface area and hence the cooling surface. (**Figure 2.3b** illustrates what this cluster looks like in the hive by using the example of an early swarm that had developed under a branch of a tree by August in the North Dublin suburbs.) Weak or unhealthy bees can struggle to maintain brood-nest temperature. Beekeepers need to be mindful of this special characteristic of the honeybee colony and keep strong colonies; it also has implications for hive floor design that will be discussed in para 3.2.

The honeybee does not reproduce in the usual way of the animal world, with a male and female of the species mating and the individual(s) raising the family, for example, as is the case with bumblebees. The need to maintain a high temperature during brood rearing, as described above, dictates that reproduction must take place by division or swarming in the case of honeybees. It allows the cluster to take advantage of scale as in the case of a winter cluster where typically the surface area per bee is 0.07 cm$^2$ compared to about 2 cm$^2$ for an individual bee[181].

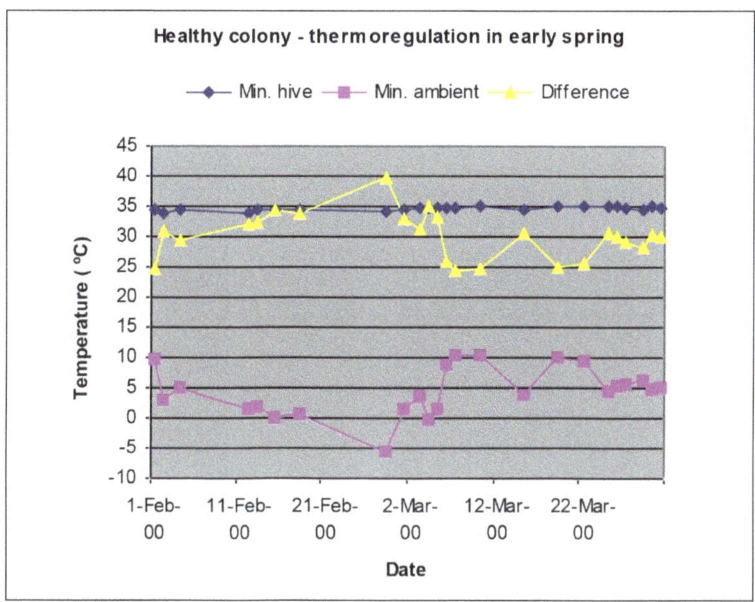

**Figure 2.3a** A typical temperature profile in the brood nest of a healthy honeybee colony in early spring. Precise thermoregulation is present even with large changes in ambient temperature.

**Figure 2.3b** Swarm settled below tree branch in North Dublin suburbs in late august. This colony would be unlikely to survive an Irish winter.

Communication is critical to the functioning and survival of the honeybee colony as a unit. It has been developed to a high degree and uses a series of methods including food sharing, scent, dance and vibration. Much of this is exercised through pheromones, secretions that act as chemical messages at the individual bee level.

## 2.4 Honeybee pollination

**Figure 2.4a** Honeybees working on white clover (L) and oilseed rape (R)

Pollination takes place when pollen is moved from the anther (male part) to the stigma (female part) of a plant of the same species. The honeybee is an important pollinator of the natural habitat as well as many food crops **(Fig. 2.4a)**. Its value as a pollinator has been recognised from the time of Aristotle. It is interesting to note that Arthur Dobbs from County Antrim (described as an Irish squire, naturalist and beekeeper) in 1750 concluded through his observations that a honeybee remains constant to one particular kind of flower on each foraging expedition[60]. This confirmed what Aristotle had concluded, although much opinion at the time was to the contrary. While the economic contribution of pollinators to the food supply is considerable, it is their contribution to the wild flora that produces the blossoms, seeds and fruit for our wildlife that is most important, as well as enhancing the beauty of the countryside **(Fig. 2.4b)**.

Besides honeybees, there are other important insect pollinator species; in Ireland 76 solitary bees **(Fig. 2.4c)**, 20 bumblebees and some 180 hoverflies. However, pollinators in Ireland and worldwide are in decline. The All-Ireland Pollinator Plan 2015-2020[148] was developed to address this decline with the aim

**Figure 2.4b** Above, hawthorn blossom before pollination, and later haws (below) showing how the countryside is enhanced in both stages.

of identifying actions to restore populations of pollinators to healthy levels. The cumulative effect of the plan and the accompanying series of sectoral guidelines will enhance the environment for honeybees (and other pollinators) through improved awareness, redressing habitat loss, increasing availability of wildflower and blossom as well as discouraging the use of chemical sprays[148].

**Figure 2.4c** Solitary bee pollinators: Top, *Osmia rufa*, a recent import for horticultural crops and below, *Andrena nigroaenea*, a fairly common pollinator in eastern and southern Ireland

Chemicals are increasingly being cited as having a major impact on the availability of healthy forage for pollinators. The major concern has been with the use of the neonicotinoids and recent research, with some exceptions, has shown it to have a predominately negative impact on bees[174,219]. The negative impacts on bumblebees and solitary bees are generally more pronounced than with honeybees, but long-term persistence of neonicoinoids in the soil is a common finding. Sub-lethal and combined pesticide exposure impact negatively on bee health[62,91,120].

## 2.5 Beekeeping regimes

There are many ways that beekeepers can manage their honeybee colonies and it is usually dictated by the scale of the beekeeping operation, but not always. Sometimes the main aim of the beekeeper is to maximise honey production, as is usually the case with commercial beekeepers (or likely for anyone who has more than 5 to 10 colonies). The beekeeping approach will generally involve some compromise with the bees' health, both short and long term. This would certainly include the commercial operations such as those in North America where large-scale migratory beekeeping is accompanied by the use of pharmaceutical products and the employment of bought-in services for package bees or queens. Of a lesser intensity is the serious hobbyist working the bees on a part time basis involving large numbers of colonies, which describes a significant proportion of the beekeepers in North America and some parts of Europe. Bought in bees (nuclei or package bees) and queens are common as is mass production of queens. These two regimes will not be given particular emphasis in this book, but reference will be made to general or specialist books as appropriate. The emphasis here will be with the remaining beekeeping regime, the small-scale beekeepers who account for the vast bulk of beekeepers in number (but not necessarily of colonies). The text will show how colonies in this category can be managed. A sustainable approach to beekeeping will also be demonstrated (para. 5.1, 5.2).

# PART 3 | The Beekeeping Set-up

There are many decisions taken by beekeepers in setting up their beekeeping activity. Getting them right will provide an ongoing benefit to the honeybee colonies, and mostly come with little extra effort or cost. They will not guarantee success but will increase the likelihood of having healthy and productive colonies. The hygiene involved in handling the colonies, as well as disinfecting and sterilizing is dealt with in Part 5, *Working with the Bees*.

As a pre-requisite, all beekeepers should be trained before receiving their first colony. Beekeepers have a *duty of care* to their bees and without training are unlikely to be able to handle them properly, will crush an excessive number and will be a potential nuisance to neighbours through bad siting and swarming (para. 3.5 below, *The beginner-beekeeper).*

## 3.1 Apiary

The apiary choice is critical. Do not assume that it will be located at your home place. It may be unsuitable due to congestion in an urban area, or the size and orientation of the site. Neighbours have a right to enjoy their property. Get advice from a mentor or experienced beekeeper. It may be necessary to use an out-apiary, a site some distance from home.

- locate the apiary in an area that has adequate forage (pollen and nectar sources) to permit the maintenance of healthy, strong and productive colonies. The presence of lots of trees will not be relevant if they are neither nectar nor pollen bearing.
- The apiary should have good access, be well fenced and preferably be out of public view.
- ensure that the forage in the catchment area is sufficient for the number of colonies. Don't overpopulate with hives and respect the rights of existing

beekeepers in the area. Initially, local beekeeping knowledge will help but later the availability of forage in practice will confirm the choice.

**Figure 3.1a** Choose sunny, sheltered location in an area with a good all-year-round supply of forage.

- select sunny sites free from overhanging trees yet sheltered from prevailing winds **(Fig. 3.1a).**
- use raised hive stands to enable good air circulation (and also to reduce back strain). The stands should be staggered and kept apart (1.2 to 1.5 m) to reduce drifting and provide space for manipulating colonies.
- keep colonies in weather-proof hives (para. 3.3) and number all hives and supers. It is desirable that hives face in a generally southerly direction[183].
- keep a dedicated beekeeper's toolbox and regularly clean the contents (para. 3.4**)**

## **3.2** Key hardware considerations

Some of the key set-up decisions deal with the choice of beekeeping hardware. Other key choices, such as the honeybee strain, sustainability, raising queens will be covered in Part 5.

## Hive types

Beekeepers should not lose sight of the fact that the evolution of the honeybee, as a cavity dweller, would have been in rotten hollows within tree trunks **(Fig. 3.2a).** The nest was typically high up in the tree, with a very small entrance (4 cm dia.) and was surrounded by 6 to 10 cm of rotten wood[185]. The hive was therefore highly insulated and provided a thermally efficient arrangement.

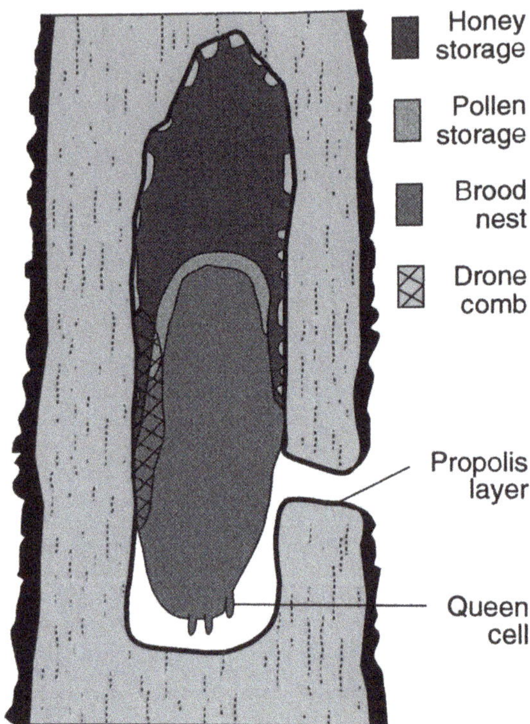


**Figure 3.2a** Schematic of a vertical section through a typical nest in a hollow tree (from Seeley and Morse, 1976)[185].

While feral colonies still occupy these sites, beekeeping has progressed from tree cavities to skeps **(Fig. 3.2b)** and on to the assortment of moveable frames hives that we have today. It is prudent to confine your choice of hive to those available in the locality, which allows for the flexibility of receiving colonies on the same brood frames. In any case avoid having mixed hive types. Keep it simple; pick only one standard type and it is unlikely to make much difference to the bees. Key additions to hive design in recent times are the use of open-mesh floors, and polystyrene material and these will be considered below.

**Figure 3.2b** A traditional 19th century straw skep with split-blackberry cane lapping.

In Britain and Ireland, the National hive still appears to be the most popular, whereas the Langstroth is the most popular in most of the rest of the world. If we had had to start afresh in Britain and Ireland with no existing stock of WBC or CDB hives (with British Standard frames), would the larger Modified Commercial hive (+50%) or the Langstroth hive (+30%) be the most popular today? (A Langstroth brood box is similar in comb area to that of feral colonies of European honeybees in the northern US states[183]). Both National and Commercial are square hives, with similar external dimensions, and this allows a National super to be used in place of a Commercial one if weight is a problem. Both hives have traditionally had bottom bee space. Dummy boards are less common with National hives, presumably because beekeepers do not want to further reduce the already limited brood space. Due to this reduced brood space beekeepers sometimes work their bees on two brood boxes. Beginners can find it easier to hold the lighter brood frames and longer lugs of the National hive. On the other hand, the Modified Commercial is relatively deep and has also square internal dimension. This allows the brood box to better approximate to a cube or sphere, giving it a higher ratio of volume to surface area. This would enhance thermoregulation, important in cool, damp climates.

There is also a niche interest where some beekeepers wish to use comb that the bees have built in the natural way without resorting to foundation. This was the case with skep beekeeping and present-day alternatives are available such as the Top-Bar **(Fig. 3.2c)** and Warre hives that use only the top bar part of a frame.

Figure 3.2c A top-bar hive where the bees draw out comb by fixing it to a top lath of wood. It mimics the traditional log hive used in Africa.

Hives have been traditionally made from wood either Western Red Cedar or pine. Hives made from synthetic materials such as polystyrene and polyurethane have become available in the recent past. They offer lightness and good insulation qualities. However, they tend to be less durable than wood particularly working in full hive mode with supers and can be prone to deterioration due to UV radiation. The brood boxes need to be cleaned/sterilized with washing soda or bleach, sodium hypochlorite as a blow torch cannot be used (para. 5.4). These are more commonly used in the nucleus mode, but the beekeeper needs to ensure that brood frames are interchangeable with their other equipment.

## Open-mesh floors

A widespread use of mesh floors in hives occurred with the arrival of *Varroa* mites. Initially a screen introduced between the brood box and the solid floor enabled natural mite drops to be measured and these were used as a proxy to estimate the seasonal mite population. This was later used as an Integrated Pest Management (IPM) step as it enabled any subsequent treatment to be given only as required to keep the mite population below the economic injury level

(EIL), namely the mite population level above which economic damage is caused. IPM took various forms, many of which were labour intensive and disruptive to colonies. They included shaking powdered sugar on brood combs to make the mites *lose their footing*[65]. Working with sustainable colonies (locally adapted) as described in paragraph 5.1 removes the need for IPM and colony treatments, however exceptionally a 'varroa-specific artificial swarm'[76] (para. 5.2) may be required.

There was evidence that a screen between the brood box and the solid floor (closed-mesh floor) can reduce the mite infestation level compared to a solid floor, but this has been shown not to be significant in most of the studies[43,69]. The mechanism appears to be a reduction in the number of dropped mites returning to the brood box. While there are claims that using an open-mesh floor (a floor with a mesh replacing the solid floor) will further contribute to mite reduction, there has been little work done on this in temperate climates[69]. There was some evidence that open-mesh floors contributed to increased brood production[99]. The reason for this is not clear. It may be that some of the anecdotal claims for the open-mesh floor in these islands is where beekeepers using the closed-mesh floors have had difficulty keeping the solid floor dry, as the bees have no access to the solid floor to remove the water. In these circumstances, using an open-mesh floor will reduce the possibility of water collecting on a solid hive floor and would thereby assist the colony to thermoregulate, particularly during winter and early spring.

As already mentioned, bees evolved as cavity dwellers in relatively small hollows in trees that were well insulated with dry, rotten wood all around and that had only one small entrance[183]. Also, the initial research on the use of open-mesh floors was undertaken in the sub-tropics of Baton Rouge in the southern United States. Recent research undertaken in Ireland questions the use of open-mesh floors in a temperate region in terms of both *Varroa* mite suppression and general colony health[129]. During the study, a large temperature difference (~ 19°c) was measured in early spring between the top of the solid floor of hives and ambient. With an open-mesh floor, this point in the hive would tend towards ambient, 19°c colder. The preference of *Varroa* mites for reproduction in cooler brood cells would give them an advantage, particularly in the critical drone cells[107]. Interestingly there seems to be an increasing trend of drone brood being observed higher up the brood comb, which would put the mites at a reproductive disadvantage due to a raised drone brood temperature and consequently reduced drone emerging times[118]. The study also showed that very few of the mites that had dropped had the potential to return to the brood nest. The use of open-mesh floors also exposes colonies in the temperate regions to greater thermal risk during the spring build-up period.

Where beekeepers are no longer counting mite falls, a mesh insert in a solid floor is redundant and it is suggested that the traditional solid floor be used. However, where beekeepers use open mesh floors, keeping the insert below the open-mesh floor during the winter/spring period in temperate climates would be prudent, and ensure the insert does not get wet. Inserts should provide good insulation, for example, wood ideally not less than 10 mm thick. Also reduced entrance blocks should be used to assist colony thermoregulation during the winter/spring period.

## Small-cell combs

The use of small-cell brood comb has been advocated for a long time by "natural" beekeepers, particularly those in the United States. The belief was that beekeepers moved from the small to the standard large-cell comb towards the end of the 19th century, following the trend in agriculture at the time that "big is good"[133]. The move away from natural comb sizes was considered to be associated with bee diseases particularly the parasitic tracheal and *Varroa* mites. The author's involvement with small-cell combs started in 2001 with an investigation into a possible link with tracheal-mite infestation. However, there was no evidence that small-cell size had any influence on the susceptibility of honeybees to tracheal mite infestation[132].

It was also thought that large cells contributed to *Varroa* mite infestation. Reducing the cell size to the natural size may restrict the movement of the male mite as it moves from the top of the brood cell to its feeding site, thereby reducing mite reproduction and hence mite infestation. The native Irish dark bee, *Apis mellifera mellifera,* was able to draw out small sized comb with little difficulty in contrast to some of the other subspecies in Europe. It was established that reducing cell size from a nominal cell width of 5.5 to 5.0 mm increased the "fill factor" (ratio of thorax width to cell width) in the cell from 73 to 79%[133]. The change in fill factor was corroborated by work in the United States published in 2011[184]. However, recent experiments to establish a corresponding reduction in *Varroa* mite infestation with small-cell brood comb in Ireland[45] and the United States[184] did not establish any advantage. Unfortunately, the use of small cells, which would have been a very attractive string to the mite management bow, proved of no benefit in containing the mite population.

## 3.3 Hive and stand arrangements

### Hives

All modern hives have a similar arrangement of components; a floor base upon which rests the brood box containing frames of wax comb, a queen excluder, honey supers, crown board and a roof on top. Most beekeepers choose to align the frames parallel to the front of the hive, the 'warm' way, but they may be aligned at right-angles, the 'cold' way. The warm way, as the name implies, will tend to have a reduced cooling effect on the colony and will have a brood spread that is generally symmetrical about the centre line of the brood comb.

It is essential that hives be properly maintained. In temperate climates, with high rainfall, keeping bees in weatherproof hives is critical for healthy bees. Colonies will find it difficult to maintain brood-nest temperature in wet hive conditions and the resulting stress can bring on diseases such as nosema and a slow spring build-up. The author has remarked over the years how much better the relative health and productivity of colonies are in an open hay shed in a high-rainfall region compared to those not under cover nearby **(Fig. 3.3e)**. In these dry hive conditions, earwigs will be common and slugs/woodlice scarce.

Figure 3.3a A double-nucleus hive made by putting a partition board in the centre of a standard hive. The cross-section (inset) shows five frames plus a dummy board on each side of the central partition and entrances to front and rear. The compact brood nest illustrated is a thermally efficient arrangement for over-wintering small colonies.

A double-nucleus hive is an arrangement that is particularly suitable for over-wintering small colonies. It is made up from a standard hive with a dividing board in the centre of the brood box that can be slid in between two shallow guides on each side and fitted flush with the bottom of the hive floor. When re-used as a full hive the shallow runners will not interfere with the movement of the frames.  The arrangement is completed by securing a closed entrance block to the front and making two-half crown boards over the five-frame nuclei. The floor can be securely fixed to the brood box with broad (50 mm) tape and two small entrance holes (~ 10 mm) drilled, one to the front and another to the rear. This arrangement is compact and is shown in **Figure 3.3a** with a super on top to provide headroom during feeding. The inset shows a cross-section of the brood box and the compact brood nest, which parallels a single colony in a full hive. With two balanced nuclei in place, the heat loss is cancelled across the division board giving the nuclei the thermal efficiency of a much larger colony. Colonies in a double-nucleus box over-winter well and build up quickly in the spring. The hives are physically much more stable than single nuclei on a hive stand during windy winter weather. They are particularly suited to a small-scale beekeeper who may wish to over-winter a few nuclei but wants the flexibility to revert to full hive operation later with the same equipment.

## Stands

**Figure 3.3b** A CDB hive from the late 19th century. The design included legs and a landing board and was usually placed on a plinth that allowed good air circulation.

Hive stands fulfill many needs. They raise hives off the ground and reduce the level of stooping when manipulating or moving colonies. Stands can make it easier to lift or put straps on hives and can provide access to enable hives to be weighed (para. 5.5). Probably the most critical function that they perform is reducing the level of dampness around the colony by raising the hives well clear of the ground and surrounding vegetation. It is amazing how many beekeepers use whatever is at hand even though proper stands are inexpensive (and easily made) and when written-off over the life of the hive amount to a fairly small annual charge. It is common to see hives resting on concrete blocks, old milk crates and even used tyres. As a general rule stands should be set-up reasonably level.

Beekeepers operating on a large scale may have special needs where they are constantly moving hives over wide areas for pollination or to follow the forage. The economics will also be a factor. The small-scale beekeeper with only a few hives will, however, not have these constraints and should aim to use those that are best suited to the bees. Stands, waterproof hives with the proper siting of apiaries (para. 3.1) are three critical steps in ensuring healthy bees. They will not guarantee healthy bees, but if beekeepers do not get these steps right, they will not be giving their bees the best start on the road to health.

During the move away from skep beekeeping in the second half of the nineteenth century, it was common for the earliest hives to have a combined floor and stand. The legs of these hives tended to be short. However, this was often compensated for by standing the hive on a concrete plinth that allowed for good air circulation. Examples of these hives are the WBC in Britain and the CDB from an Irish design of the 1890s (**Fig. 3.3b**). These hives, while a lot prettier than those of the present day, required a lot of stooping during manipulation. They used British Standard frames (National), were double walled and are still used today as *warm* hives particularly suitable for producing section honey.

Later hives required the beekeeper to provide a stand, and this usually resulted in many and varied stand types. However, some suitable types are in use and a sample of a few different concepts in stands are shown.

Portable double stand **(Fig.3.3c)**. This allows for good ground clearance and with the floor of the hive being almost 600 mm off the ground, ensures that the brood box is at a comfortable height for working. It has retractable legs that make it readily portable and at a length of 1.8m, can accommodate up to two hives and a nucleus. Dampness and vegetation growth in the immediate vicinity is reduced with the use of horticultural carpet around the stand.

Euro-pallet double stand (**Fig. 3.3d**). Euro-pallets provide a cheap and effective double stand. One or two rows of 150 x 220 mm concrete blocks on

**Figure 3.3c** A portable double stand ('Clonroche' stand), with retractable legs and a skirt of horticultural carpet.

**Figure 3.3d** A Euro-pallet double stand with good access from below and a skirt of horticultural carpet.

horticultural carpet provide a suitable base. The pallets are open on the top, which allow good access below for lifting or weighing and can be put one on top of the other to adjust the height.

Split railway sleeper stand (**Fig. 3.3e**). Railway sleepers can be split and bolted to 600 x 600 mm concrete flags. This arrangement provides a robust support, and the lower height makes it particularly suitable in situations where a large number of supers are the norm.

Figure 3.3e A split railway-sleeper stand provides a robust support for multiple supers. This colony is in an open hay shed.

## 3.4 The beekeeper's toolbox

The selection and proper maintenance of the tools to manipulate the colonies can affect the treatment of the bees and the general hygiene achieved. This is also true for the container in which the tools are kept, but this is often neglected.

Many beekeepers do not have a dedicated toolbox. Tools are taken to the apiary in some amazing combinations of containers: cardboard boxes and even shopping baskets. At the end of the day, the choice of tools and container is largely a personal issue. However, having the correct tools and having them readily accessible can reduce colony disruption and lessen the time to undertake a manipulation. There are some practices that are useful and may be helpful to others, particularly beginners (**Fig. 3.4a**).

Figure 3.4a A dedicated toolbox, showing smoker and quench tin, hive tool (strapped to smoker), scrapings box, hive records (blue folder), weigh scales, etc, and a hinged lid.

The toolbox should be large enough to hold at least the minimum number of items without creating a "pile" of things at the bottom. Small items that can reasonably be held in the pockets of the bee suit should not be left to clutter the box. For example, one breast pocket of the bee suit could hold the queen press-in cage in a block of polystyrene and blunted clipping scissors pushed into the same block, while the other breast pocket could have matchboxes for taking bee samples or lined with cotton wool to take the occasional queen cell. Marker pens, plunger marking cage and lighters can be held in trouser pockets, with gloves, when not in use, pushed in on top to keep them in place. Having these items on your person means that you do not need to lift your eyes off the comb during a manipulation to use them. Keep things simple and avoid the latest gimmicks. Occasionally one comes across beekeepers arriving in the apiary with all sorts of bits and pieces hanging out of them. It is, however, important to have a hive tool that you are comfortable with and not too long; excessive leverage can be exerted with long tools damaging frames and corner joints of hives and allowing water to enter.

The smoker, hive tool, brush, scraper, hive straps, queen-introduction cages, spare smoker fuel and hive records (and weighing scales if used) can be kept in the toolbox. Having the hive tool pushed into a strap on the back of the smoker when in the box **(Fig. 3.4a),** allows them to be treated as a pair and thereby reducing the possibility of arriving on site without a hive tool. The strap can be a strip of metal from a bean tin, screwed onto the back of the smoker, and can be transferred to a new smoker when the time comes for a replacement. Using a bean tin to quench fuel such as hessian reduces the risk of re-kindling on the way home in the car. The toolbox must be made from material that allows it to be cleaned out and disinfected on a regular basis. A strong box can even double up as a stool.

In addition to the toolbox, the beekeeper may use a container with a lid, for collecting brace comb, cut-out queen cells, propolis and other bits and pieces that should not be left in the apiary. Hygiene demands a tidy apiary. Having a straight-sided container that can be pushed against the hive means that few items drop on the ground **(Fig. 3.4a)**.

## **3.5** The beginner-beekeeper

If beginner-beekeepers do not start off on the right foot they can develop bad practices that will be very hard to shake off.

Beginners are in a delicate state. Like most of us, they start with no previous knowledge of the insect world. In this situation beginners' courses are an essential first step where the old-craft skills can be supplemented with an understanding of the nature of the honeybee[141]. It is good to remember that insects (and honeybees) are strange animals, and if we started out with the same level of ignorance in raising our children, there would be a lot of child neglect and mortality around! We all have a *duty of care* to our bees and if we don't try to understand their needs, treat them properly and be able to relate to their behaviour and maladies, we really should not be in charge of them.

Nowadays there are so many sources of information on beekeeping matters available to the beginner that if they try to take it all in, they are in danger of becoming very confused. One good general textbook on beekeeping should be the main point of reference for the beginner in the early years. Regardless of the amount of beekeeping knowledge that the beginner has accumulated, at the level of manipulating the colony it is important that the beginner, and all beekeepers for that matter, should keep it simple. Only properly evaluated practices should be used. It is disappointing to witness, from time to time, beekeepers of many years standing suggesting spurious techniques that unfortunately mislead the keen beginner. Bees are survivors and can adapt to overcome most bad practices, but this comes at a cost.

Using experienced beekeepers to mentor is an effective way of bedding down the techniques taught in the beginner course. However, the mentor must be suitable. Don't put a beginner under a mentor with little energy; the beginner may never get started. Moreover, don't use mentors with strange, quirky practices; the beginner may never recover. The beginner should use standard equipment and should not get caught up in the most recent fads. The mentor can assist the beginner in all of the key set-up tasks, including selection of the apiary, hives and equipment, sourcing the bees, as well as assisting the beginner with key colony manipulations.

A major factor in the transmission of diseases to an area is the importation of colonies from outside. To minimize disease transmission and to nurture a co-operative approach, beekeeping associations can source bees from within their own area. Members working together can raise queens/nuclei from existing bee stocks in the area and make them available to members (para. 5.2). Furthermore, established beekeepers may wish to provide an occasional nucleus to help beginners to get started. The beginner in turn could then be required, say within two years of receiving their bees, to make a nucleus of bees available to the association for distribution to "new" beginners. In time returning nuclei from "old" beginners could be the main source of new colonies for beginners.

Mentoring for one year only means beginners must realize that they should strive to be self-sufficient within this period. This avoids the situation where beginners take advantage of the available help over an extended period, and it also frees-up the scarce mentor resource for new beginners in the following year. In some cases, "beginner apiaries" can provide a valuable service. They allow the beginner time to find an out-apiary or to build up skills before considering bringing bees into their back garden, which is especially important for beginners in urban areas. These beginner-apiaries can host several beginners, each working their own colony, and a mentor who organizes site visits and is available to assist them **(Fig. 3.5a)**. This approach allows a number of beginners to be mentored, and the group approach in effect exposes each to more than one colony-year of experience.

The relationship that builds up between mentor and beginner in this first year usually endures and should provide the beginner with a valuable source of wisdom and advice into the future.

**Figure 3.5a** Mentoring in a beginner-apiary

# PART 4 | Bee diseases and colony disorders

## `4.0` A note on honeybee diseases

Honeybee diseases are caused by parasites, living organisms that find the honeybee a suitable host in which to undertake their life cycles. Many of the parasites are specific or obligate, which means that they can only exist in or on the honeybee. These parasites can be microparasites, such as bacteria or microsporidia, and are often referred to as pathogens, whereas the larger ones such as mites are classified as macroparasites.

The honeybee colony can offer a favourable habitat for parasites. The habitat has shelter, food, regulated temperature and humidity and perennial hosts in large numbers and in close proximity. However, the parasite's life cycle must match or adapt to that of the honeybee if it is to be able to use it as a host[180]. Furthermore, the honeybee offers an effective transmission system to allow parasites to invade new colonies. Horizontal transmission (between bees of the same generation) by drifting/robbing and vertical transmission through swarming exploit normal honeybee behaviour[84].

The level of host injury will depend on host susceptibility to the parasite and the virulence of the parasite itself. The virulence of a parasite is its capacity to harm the host whereas the susceptibility of a host is the likelihood of it being infected/infested by the parasite (disease). Microparasites, such as viruses and microsporidia, are normally referred to as *infecting* colonies whereas macroparasites, such as mites, are referred to as *infesting* the colony. A measure of the degree of infection or infestation is i) *prevalence,* which is measured as the proportion of the total colony population that has the disease and is usually expressed as a percentage, and ii) parasite *load or intensity,* which is the average number of parasites present in the population of diseased bees and is expressed as the number of mites or spores per bee.

The parasite depends on the honeybee host for its existence and receives its nourishment (resources) from the bee (**Fig. 4.0**). A parasite with low virulence will fail to develop/or die in a host with a high level of resistance (low susceptibility) to it. On the other hand, if the parasite is of high virulence and the host is highly susceptible (has low resistance) to it, the host will die and the parasite will die with it. In nature, due to variability in both parasite and host, an adaptation usually takes place and a balance is struck that allows the two to co-exist. Chalkbrood and tracheal mites are examples where stable relationships have been established.

In the case of an exotic parasite, the initial contact between the parasite and its bee host can have a devastating effect on the host. (The model in **Figure 4.0** can also be applied to human diseases such as the exotic coronavirus COVID-19, that was particularly virulent when it first arrived at the beginning of 2020[143]. Each new variant will go through a process of attenuation in virulence over time) This is exemplified by the arrival of *Varroa destructor*, which had evolved with a different host, *Apis cerana*, and switched to a new host, *Apis mellifera*, that had little resistance to it. In the original host, the mite only reproduces in the drone brood, but on its new European host, *Apis mellifera*, the *Varroa* mite reproduces in both worker and drone brood resulting in high mite populations. In the first edition of this book[128] in 2012 when dealing with the initial impact of *Varroa*, it was stated: *However, if left to themselves to evolve and adapt without the "help" of beekeepers treating their bees against the threat of the parasite, a balance would be struck but at a high cost in the short-term in terms of colony mortality.* In 2012 the author was in the third year of non-treatment and had to be circumspect in relation to stating any conclusions. Since then, colonies have adapted to their *Varroa* parasite without the predicted high cost in the short-term (para. 5.1, *The bees & sustainable colonies*).

The usefulness of treatments can also be short lived, as the parasite can develop resistance to the treatments. In addition, variability in the parasite can result in resistant strains of *Varroa destructor* being selected that can be dominant within a short period. Beekeepers who don't follow the manufacturer's instructions when using chemical treatments also contribute to parasite resistance.

However, the honeybee throughout its existence has been responding and evolving to the threats in its environment. Many physical defences are present. The natural cleanliness of the bee is always remarkable. Foreign matter and sick and dead bees are removed to the entrance and are usually carried and deposited a long distance from the hive. Autogrooming and bee-to-bee grooming (allogrooming) are part of a bee's normal activity. There are also many chemical defences available. The acidity of honey, its hygroscopic (moisture absorbing) properties and the natural presence of hydrogen peroxide all help to protect against microparasites such as bacteria. Honey, whether added to pollen to make it stick to the bee's pollen baskets or to glaze it in the cell for winter storage, provides a natural preservative for pollen.

The existence of propolis, while much maligned by many beekeepers, is an important addition to the hive. It is seen as a possible element in social immunity in honeybees[189], providing a complex resin with wide antimicrobial properties. There is evidence that propolis inhibits the growth of the AFB bacterium[28,10] and reduces the reproductive ability of *Varroa* mites[189]. Propolis is used to embalm large hive intruders, such as mice and slugs, and is known to encapsulate the parasitic small hive beetle[151].

Adequate nutrition is critical for a colony's growth, development and health. With changing agricultural land use, we cannot take it for granted that a colony's nutritional needs are being catered for, particularly in relation to pollen. Poor nutrition may reduce the ability of bees to withstand the many diseases present in today's world and have serious consequences for the health of the colony[150].

Polyandry, the multiple mating of honeybees, has given bees an increased ability to tolerate diseases, but it is only effective if a good spread of genes is available[199]. Intensive queen rearing on a global scale has diminished the mix of honeybee genes. This has been exacerbated by the arrival of the *Varroa* mite, which has reduced feral (wild) colony numbers and further diminished the gene pool. Reducing the local gene pool may seriously weaken the bee's ability to withstand the onslaught of disease.

When more than one parasite or pathogen is present, there is generally a synergistic effect where low levels of parasites, which on their own would not be of much concern, can have a traumatic effect when occurring together. Well-known examples are the combination of *Nosema* with tracheal mites[12] and also of *Varroa* with tracheal mites[61]. Parasites in the presence of pesticides can also have a similar effect[208].

There is a large diversity of organisms that are associated with honeybees[92]. While most of them are beneficial, some are very harmful. Essentially four parasite/pathogen groups are responsible for honeybee diseases. These can be categorised as i) bacteria, ii) fungi/microsporidia, iii) viruses and iv) mites/beetles.

## Bacteria

These are mostly single-celled organisms. The cells are simple structure with no nucleus to contain the genetic material and are referred to as prokaryotes[6]. They are prolific in numbers and types, but few of them cause disease[22]. However, although most bacteria are harmless two distinct bacteria are responsible for the well-known brood diseases American and European foul brood, the former, along with tracheal and *Varroa* mites being the most virulent parasites in the history of our European honeybees[88]. These organisms are small and a microscope magnification of x1000 magnification is required to identify them.

## Fungi/Microsporidia

**Fungi** are large in numbers and types and unlike bacteria have a nucleus and are referred to as eukaryotes. Fungi have the distinction of being the first disease diagnosed as a micro-organism when in 1834 a fungus was identified as the cause of a larval disease in silk worm[22]. A fungus causes the widespread condition known as chalkbrood. Based on recent molecular evidence, microsporidia are classified as highly specialized parasitic fungi[190].

**Microsporidia** are spore-forming organisms that have recently been classified in the fungi family. In the past, they were also classified as protozoa. Microsporidia are intracellular organisms and reproduce inside the cells of their hosts. The infective mechanism is via the mechanical injection of the germinating spore into the host cell[83]. The *Nosema* species (*Nosema apis* and *Nosema ceranae*) are microsporidia and are widespread parasites of the honeybee.

## Viruses

These are types of non-cellular organisms and are unique parasites in that they do not have a metabolism of their own[6]. They cannot stand alone and develop, as in the case of normal parasites by taking nutrients from their host. They need their host's living cells to reproduce[22]. But it is their ability to adapt quickly to changing conditions that can make them a formidable pathogen. The transmission mechanism used by the virus has a major influence on its infectivity and virulence. For example, viral particles from the *Varroa* mite, delivered into the bee host's fat body tissue can change the infectivity and virulence of an otherwise relatively benign virus. Virus particles are very small and can only be examined under an electron microscope.

## Mites / Beetles

**Mites** are classified under the order *Acari* and are highly specialized animals that have colonized all places where animals can exist[63]. The number of species may be in the same order of magnitude as insects (>1 million). They are arthropods, having segmented bodies and jointed limbs. Mites have four pairs of articulated legs, a pincer-like mouthpiece but no eyes, antennae or wings. The two well-known macroparasitic-mites of the European honeybee are the tracheal mite (*Acarapis woodi*) and the *Varroa* mite (*Varroa destructor*). The *Tropilaelaps* mite, a parasite of the giant Asian honeybee, *Apis dorsata,* although not presently found in Europe represents a future threat.

**Small hive beetles**, *Aethina tumida*, have been introduced to the US, Canada and Australia and are multiplying rapidly. Although found in southern Italy in 2016, they have not been found in other parts of Europe. As in the case of the *Tropilaelaps* mite, the small hive beetle poses a future threat to beekeeping.

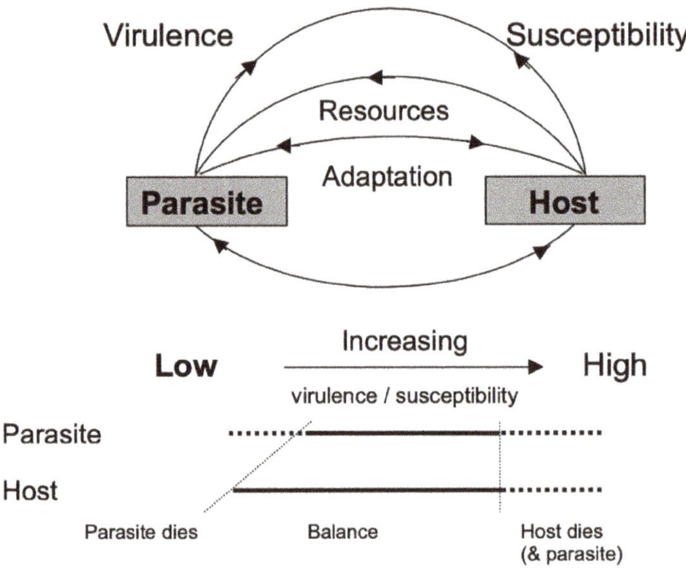


**Figure 4.0.** Schematic of parasite-host interaction. The diagram shows how a balance (or tolerance) can arise due to variability in parasite virulence and host susceptibility (from McMullan, 2012)[128]. The initial infestation of colonies by *Varroa* mites will typically start on the right with colonies dying due to high mite virulence and low honeybee resistance (high susceptibility). Over time, in untreated colonies, reducing mite virulence and increasing honey bee resistance will be selected resulting in a balanced relationship.

# Bacteria

## 4.1 American foul brood (AFB)

### Cause

American foul brood (AFB) was identified and named in 1906[213]. It is caused by the spore-forming bacterium[214], now named *Paenibacillus larvae* after many re-classifications[102,90]. When larvae are given food containing AFB spores, the bacteria proliferate in the gut and infect the tissue[224]. Larvae die after the cell is sealed and then melt down into a thick and sticky mass that later dries to form a scale on the cell wall. The scale is not moveable by the bees.

### Signs/Diagnosis

Being familiar with healthy brood allows identification by comparison (**Fig. 4.1a**). The bees should be shaken or brushed off combs to enable a clear inspection. The signs can be diverse:

- *open* brood usually has no signs
- signs on *sealed* brood include[98]
    - a few darkened cappings at an early stage (**Fig. 4.1b**)
    - cappings moist, dark, greasy, sunken; also perforated cappings
    - "pepper pot" appearance in comb
- if a matchstick is pushed through a sunken capping and twisted, the brownish contents will rope up to 2 cm (one inch) when the matchstick is slowly withdrawn (**Fig. 4.1c**)
- contents finally dry to a very dark scale at the lower end of the cell
- field test for AFB is the "matchstick" test above, and/or presence of immovable scale. Confirmation should be made by microscopic examination of the comb in the laboratory. Lateral flow devices (LFD), developed in conjunction with CSL (UK) are also used to confirm the presence of the bacteria[74]

### Virulence/Spread

AFB is present in almost all countries where honeybees are found. It is a highly virulent disease and can kill colonies in one season, and spores remain infectious for more than 35 years[88]. Young larvae are particularly vulnerable[33]. The principal means of spreading the bacteria are through[98]:

- bees having access to infected combs or feeding infected honey
- robbing/drifting
- bees obtained from dubious source

- use of second-hand infected equipment
- comb exchange between colonies, where health status is unknown

## Notification/Treatment

AFB is a disease that must be notified to the beekeeping authority in most countries. Beekeepers should be aware of the current *National Disease Policy* (para. 4.17).

- If AFB is suspected, the comb containing diseased brood should besent to the diagnostic laboratory for testing using microscopic, PCR or antibody techniques, such as the LFD (UK) method[156]. The sample should be loosely wrapped in a newspaper, not squashed (para. 5.7). All other colonies in the apiary should be examined.
  Note: If the field diagnosis is convincing, the beekeeper should discuss destroying the affected colonies with the approving lab pending official confirmation, subject to national practice.
- There is no approved chemical treatment for AFB; the practice is to destroy colony. The hive should be sealed up to deter robbing when flying has stopped. Open-mesh floors should be completely sealed. All confirmed colonies should be smothered by pouring a half litre of petrol into the feedhole of the crown board. It may be necessary to do this work under the supervision of the national authority.
- A hole (900 x 900 mm, 600 mm deep) should be dug, and the inside of the hive scraped. All bees, scrapings, brood and super frames (all combs and honey) should be burned in the hole (**Fig. 4.1d**). The inside of all wooden parts should be blow-torched to coffee brown, with emphasis on crevices. All tools and gloves should be disinfected/sterilized and the bee suits washed (para. 5.4).
- Early detection and swift action are necessary. To control and eliminate AFB, it is essential that a collective responsibility be adopted by all beekeepers in the vicinity of the outbreak.

**Figure 4.1a** Healthy sealed brood showing even cappings.

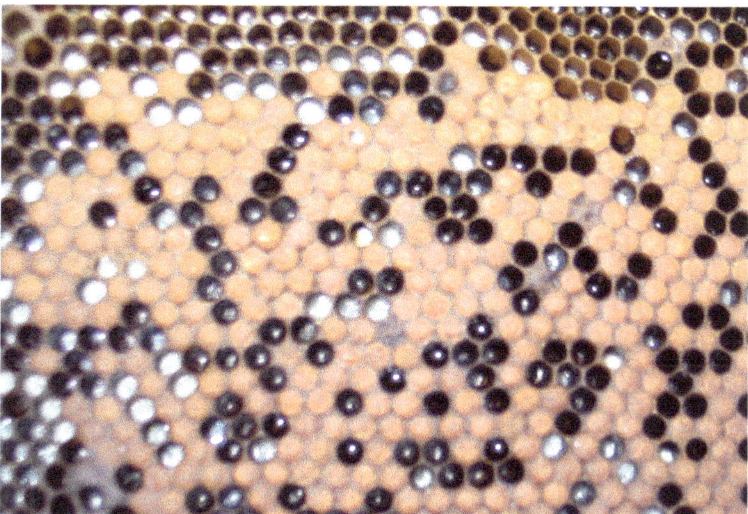

**Figure 4.1b.** A brood comb with a few darkened cells showing early signs of AFB infection. Early detection is critical to contain the spread within and between apiaries.

**Figure 4.1c** Matchstick test for "ropiness".

**Figure 4.1d** Burning the brood and super frames, bees and scrapings in a hole dug to 600 mm (2 ft.) deep. Backfill and compact soil when everything has been burned.

# 4.2 European foul brood (EFB)

## Cause

European foul brood (EFB) was first isolated in 1880s[42], was named in 1912[215] and is caused by the bacterium *Melissococcus plutonius*[215,11,24,206]. It is found in *Apis mellifera* and other species in the *Apis* genus including *Apis cerana*[18]. When larvae ingest the bacterium, it multiplies within the larval gut and competes for the available brood food. When brood food supply is restricted there is evidence that larvae may die of starvation[21].

## Signs/Diagnosis

As in the case of AFB, being familiar with healthy brood, both in the open and capped cells, permits identification by comparison (**Fig. 4.2a**). The bees should be shaken or brushed off combs for clear inspection. The signs can be diverse and can fluctuate widely over time[22]:

- usually only *open* brood are affected
- brood pattern may appear patchy and random where nurse bees have removed infected larvae (**Fig. 4.2b**)
- larvae are displaced from normal coiled position in the cell (**Fig. 4.2c**)
- larval gut initially a light creamy-yellow colour, loses segmentation and melts into a liquid mass (**Fig 4.2d**)
- contents dry to a dark brown scale that can be easily removed
- if the larvae die after the cell is sealed, the capping can have the appearance of AFB but the cell contents are not "ropey"
- an unpleasant odour may be present with higher infection. It is caused by a secondary bacterium such as *Paenibacillus alvei*
- signs can disappear spontaneously but return later
- confirmation should be made by microscopic examination of the comb in the laboratory. Molecular tools or LFD kits developed by CSL (UK) can be used to confirm the presence of the bacterium[204]

## Virulence/Spread

The prevalence of EFB varies widely within and between countries. It is currently in all major beekeeping countries with the exception of New Zealand. In some European countries the disease is increasing rapidly[170]. The major source of infection is through contaminated food, where infected larvae that survive defecate in the cell before pupation[13]. The incidence of the disease tends to be seasonal and appears to be linked to colony stress conditions, such as poor honey flow or low nurse to young larvae ratio in spring. Infection can survive in old combs for a number of years. As with AFB infestation, the principal means of spreading the bacteria are through:

- bees having access to (or fed) infected honey or combs
- robbing/drifting
- bees obtained from dubious source
- "stressing" bees through unnecessary handling or movement
- use of second-hand infected equipment
- exchange of combs between colonies

## Notification/Treatment

EFB is a disease that must be notified to the beekeeping authority in most countries. Beekeepers should be aware of the current *National Disease Policy* (para.4.17).

- If EFB is suspected, the comb containing diseased brood should be sent to the diagnostic laboratory for testing using microscopic, PCR or antibody techniques, such as LFD kits. The sample should be loosely wrapped in a newspaper, not squashed (para. 5.7). All other colonies in the apiary should be tested.
- Confirmed cases require specific and swift action:
  - weak (less than 5 frames of brood in the main season) or high-infection* colonies: destroy and manage as for American foul brood (para. 4.1).
  - "low-infection" colonies; use a "shook swarm"[74,211]. The beekeeper may also opt to destroy the colony as with AFB.

* where there are at least a few cells affected on every brood frame. An unpleasant *odour* may indicate a long-standing infection.

Note: EFB policy is currently under review in some countries[203].

- Disinfect/sterilize tools and gloves, and wash bee suit (para. 5.4).
- Early detection and swift action are necessary. It is essential that a collective responsibility be adopted by all beekeepers in the vicinity of the outbreak.

**Figure 4.2a** Healthy open brood; consistent pearly-white colour, lying in a 'C' shape in the cell with segmentation rings showing along the length of the larva

**Figure 4.2b** A patchy brood pattern showing many open cells with brood in unnatural positions

**Figure 4.2c** Brood comb of EFB infected brood showing contorted larvae.

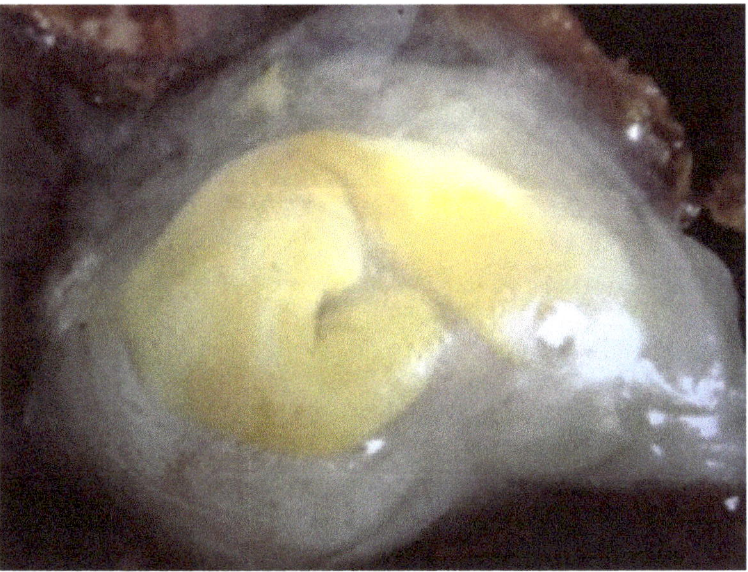

**Figure 4.2d** An EFB infected larva showing light yellow gut.

## Fungi / Microsporidia

## **4.3** Chalkbrood (*Ascosphaera apis*)

### Cause

Chalkbrood was recognised in the early 1900s[122] and is caused by the fungus *Ascosphaera apis*[192,193]. Larvae of all three castes can become infected when they ingest spore-infected food and are vulnerable up to 4 days old[20,125]. The spores germinate in the larval gut[22], and larvae reduce food consumption and eventually stop eating[202]. The fungus penetrates the gut wall and grows throughout the body cavity. Eventually the larvae form mummies that can be white or grey/black. A wide range of *A. apis* strains has been recorded[93].

### Signs/Diagnosis

Chalkbrood signs are present in most colonies. It is readily identified through the presence of white, grey/black mummies (**Fig. 4.3a.**)[188]:

- in sealed or unsealed brood cells (**Fig. 4.3b**). The mummies will rattle in the cells when tapped or shaken
- on the hive floor or at the entrance
- on the ground outside the hive

The distinct signs of chalkbrood normally mean that laboratory diagnosis is unnecessary.

### Virulence / Spread

Chalkbrood is widespread in most temperate parts of the world and where it is established, such as Europe, it is generally not considered a serious disease. It is currently found in all major beekeeping countries and a wide variation in virulence has been detected in the different strains of the fungus[93]. There is also evidence that honeybee susceptibility is genetic and is determined by the hygienic ability of the bees to detect and remove the affected brood[194]. Colonies under thermal stress in early spring or weak colonies can exhibit the condition[20,78]. Hence it is common in:

- small colonies in starter or mating nuclei
- drone brood, as they are often at the periphery of the brood nest
- weak or diseased colonies

The fungal spores are present on all surfaces within hives and can be infective for many years[22]. They can be spread during colony manipulations by the beekeeper[78]:

- moving combs between colonies and
- using infected equipment.

However, as the disease is widespread, the action of the beekeeper will only be to speed up the eventual fungal transfer.

## Treatment

Beekeepers should be aware of current *National Disease Policy* (para. 4.17).

There are no specific treatments available. The aim of the beekeeper should be to avoid the conditions that allow chalkbrood to develop through:

- keeping strong colonies
- using weather-proof hives that are appropriately ventilated
- giving attention to apiary location, colony siting and use of proper hive stands

In general, the application of good management will minimize the condition.

In the event of a severe incidence of chalkbrood and where appropriate management has been used:

- re-queening with a queen from a disease-free colony and
- transferring the colony onto new comb should be considered

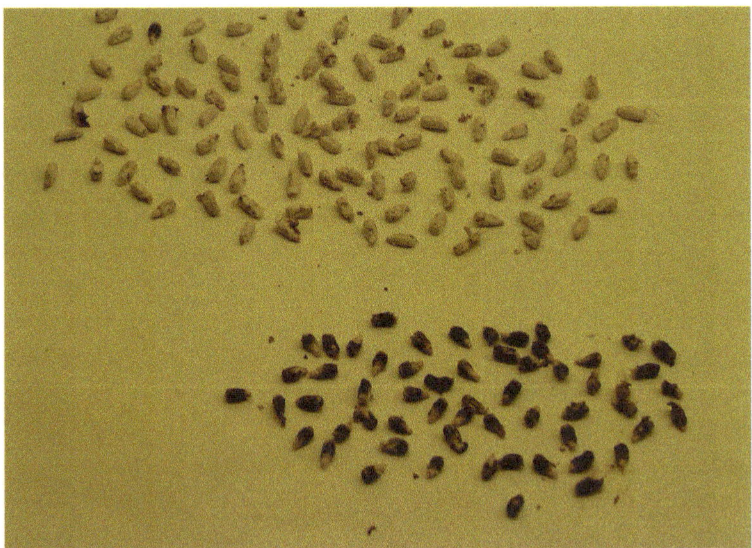

**Figure 4.3a** White and grey/black chalkbrood mummies removed from a highly infected brood comb

**Figure 4.3b** Chalkbrood visible as white chalky mummies in the brood cells

## 4.4 Nosema (*Nosema* spp.)

### Cause

Nosema (*nosemosis*) is an adult bee disease and can be caused by two microsporidia that are single-celled parasites of the honeybee gut[83]. They have recently been identified as specialised parasitic fungi[190]. *Nosema apis* Zander[225] was first identified in 1907. A new species, *Nosema ceranae* Fries[86], was identified in 1994 in *Apis cerana* and for the first time in the European honeybee, *Apis mellifera,* in Europe in 2005[103]. It was first found in Ireland in 2008. Nosema prevents proper food digestion, diminishes the hypopharyngeal glands (brood food) and reduces life span. While the pathology of *N. ceranae* is still not fully understood, it is considered a serious health threat[160] and is now globally widespread[112].

### Signs/Diagnosis

In keeping with most adult diseases, the bees show no specific outward signs:
- there is usually a slow colony build up in the spring with reduced honey production and winter survival
- *N. apis*: dysentery is generally present[22] in acute cases and sick/dead bees are found outside the hive
- *N. ceranae*: dysentery is normally absent[103] and bees are likely to die away from the hive
- the ventriculus/midgut in acute cases may be soft and swollen, with a white appearance[82] (**Fig. 4.4a**)
- confirmation can only be by microscopic examination (**Fig. 4.4b**) The *N. apis* spores are somewhat larger than those of *N. ceranae,* but it is difficult to use this to differentiate between them using a light microscope. They can be distinguished using an electron microscope or molecular-biology techniques, such as PCR–based genetic analysis[112].

### Virulence/Spread

*N. apis* is globally widespread, usually not fatal but serious when combined with other diseases. In general, it is a benign parasite that can establish itself under favourable conditions, such as low temperature stress or the presence of other parasites. Although the virulence of *N. ceranae* is still unclear, in temperate climates it can be considered as similar to *N. apis*[217,,79]. There is evidence that *N. ceranae* has a more even seasonal virulence while *N. apis* tends to reduce in virulence going into the summer. However, in warm climates *N. ceranae* appears to be a major health threat compared to *N. apis.*[144]. *N. ceranae* is becoming dominant over *N. apis* in most areas of Europe[83]. There are reports that relationships exist between *Nosema* spp. and *black queen cell virus* and queen supersedure[87,4], as well as between *N. ceranae* and CCD[159] and sublethal doses of pesticides[209].

*Nosema* is spread by spores (normally acquired when bees clean up infected combs), food sharing (trophallaxis) or grooming. Key measures that will minimise *N. apis* and assuming a similar response in *N. ceranae,* are:

- keeping strong colonies
- always using hives that are weather-proof
- avoiding crushing bees during colony manipulations
- replacing combs

## Treatment

Beekeepers should be aware of the current *National Disease Policy* (para. 4.17). The antibiotic fumagillin (Fumidil-B®) was the traditional method of treating *Nosema*. However, it is no longer available as an approved chemical in many countries, including those of the EU. This will put greater emphasis on the use of good bee husbandry. On the assumption that *N. ceranae* has a similar impact to *N. apis* in temperate climates:

- use good hygiene management. In particular build up colony strength and house in a dry hive; and this should normally rectify the condition. However, *N. ceranae* may be slower to reduce in intensity as summer approaches
- ensure that other diseases such as *Varroa* mites are kept controlled
- in severe cases, put bees onto clean combs, e.g. using a Bailey frame change or shook swarm, and use acetic acid to sterilize the old combs[75,96]

**Figure 4.4a** Highly infected ventriculus (midgut) shown in bottom photograph. (Healthy one shown in top left)

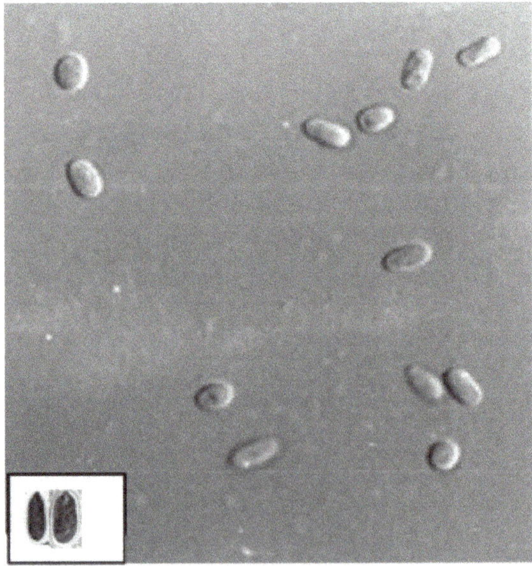

**Figure 4.4b** Spores of *Nosema apis*. Inset is the relative spore size of *Nosema apis* (to the right) and *Nosema ceranae*

## Mites / Beetles

# 4.5 Tracheal mites *(Acarapis woodi)*

### Cause

Tracheal mite infestation is also referred to as acarine or *acariosis*. The mite *Acarapis woodi* Rennie (Acari: Tarsonemidae)[165] lives within the bee's tracheae (breathing tubes) and feeds on the haemolymph (blood). It was associated with the condition that became known as "Isle of Wight disease" and that was first identified on the Isle of Wight (England) in 1904/05[1], and Ireland and Scotland in 1912. It was not until 1919 in Scotland that the mite was discovered[165]. It spread throughout Europe and later to most of the beekeeping world causing widespread deaths, particularly when colonies were first infested. Today it is not a serious condition where genetic adaptation has occurred[54]. A late arrival to Japan (2010), it has caused widespread deaths particularly to the native honeybee species *Apis cerana japonica*[175].

### Signs/Diagnosis.

There are no specific outward signs[22]:

- the mite is not visible to the naked eye: adult female ~ 0.15 mm long and ~ 0.07 mm wide
- infestation can readily be established by dissecting the tracheae of bees under low level magnification (**Fig. 4.5a**)
- infested colonies may have a slow build up in spring
- highly infested colonies may behave in a normal manner in the autumn and yet die out in early spring. Bees will leave the hive and die clinging to stalks of grass in late afternoon in early spring (**Fig. 4.5b**). Bees may vibrate or shiver their wings, but there will be an absence of body trembling or the black, shiny hairless bees, indicative of Chronic Bee Paralysis Virus (CBPV)[22].
- individual bees may have "K-wings" (**Fig. 4.5c**) and dysentery may be present at front of the hive and inside, in advanced stages
- colonies that die out in winter will generally have plenty of stores, bees clustered on a few combs and the queen present (**Fig. 4.5d**).

### Virulence/Spread

When honeybees encounter tracheal mites for the first time as a new (exotic) parasite, colony deaths are normally high. As the bee-mite relationship becomes established, mortality drops and the threat today is only to colonies where resistance has not evolved[131,52,162]. While mites migrate to young bees (<4 days old)[135] when brood is present, infestation generally increases during the winter period as old bees become infested[127]. Infestation is cyclical and colonies that do not die out in winter[14] will generally have decreasing infestation during the spring/summer only to increase again in the autumn/winter period[137]. The mechanism causing death is the inability of the bees to thermoregulate the colony[130,134,140]. While high tracheal mite infestations alone can kill colonies[134,113], even low levels of acarine in combination with other diseases can result in colony mortality[61,12].

- the mite is spread naturally through robbing and drifting[23]
- beekeepers also spread the mite by movement of infested adult bees[1]

## Treatment

Beekeepers should be aware of the current *National Disease Policy* (para. 4.17). Early detection of tracheal mites in autumn is important if the colony is to be saved[140]. It is too late to do so when colonies are dwindling, and dead bees are scattered outside the hive in early spring. The only permanent solution is to have the colony headed by a queen with resistance, e.g. from colonies that have been acarine free over a considerable period. The management approach is therefore to re-queen or to take steps to enable the colony to survive over the winter* and to re-queen in the spring[140].

- a reduction in mite population can be achieved with the essential-oil, thymol-based product, Apiguard® which is temperature sensitive, and is also effective against *Varroa* mites
- in the US vegetable patties are used in early autumn/winter to reduce the build-up of the mite population[176]. Menthol crystals are also used in the US and are temperature sensitive[71]
- winter wrapping* (putting insulating material outside of the hive in early autumn) will help the colony to survive the winter[140]. Colony can then be re-queened with a resistant bee strain in the spring
- ensure that other diseases such as *Varroa* mites are kept under control

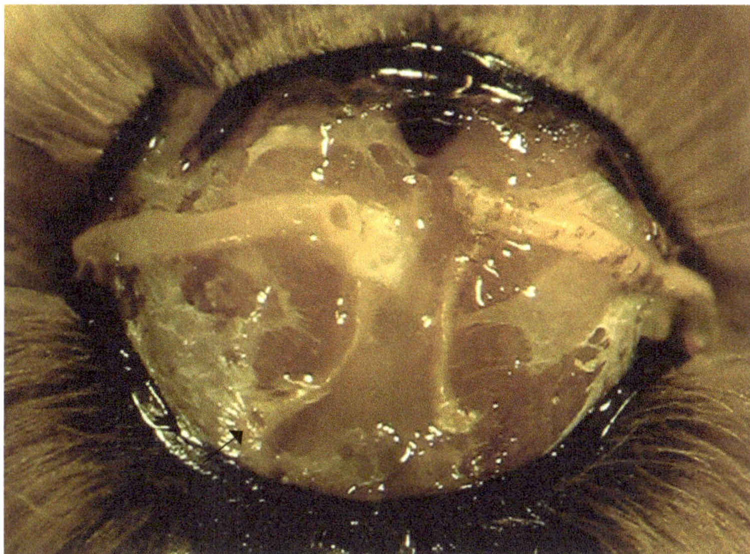

Figure 4.5a Honeybee tracheae showing an uninfested trachea to the left and infested trachea to the right. Tracheal mites are not visible to the naked eye and need x10-20 magnification.

Figure 4.5b. Honeybees huddled together, dying on the grass in late afternoon in early spring.

**Figure 4.5c** An infested bee, crawling on the ground outside the hive, displaying the "K-wing" effect. Note that the hindwing has separated from the forewing and is stuck out sideways.

**Figure 4.5d.** A typical post-mortem shows a cluster of bees, dead on the comb, with dysentery staining on their wings, and the white-spotted queen in the centre.

## **4.6** Varroa mites *(Varroa destructor)*

### Cause

Varroa (or *varroosis*) is caused by an external mite *Varroa destructor* Anderson and Trueman (Acari: Varroidae)[171]. The mite shifted host from the Asian honeybee, *Apis cerana*, to *Apis mellifera* in the first half of the last century. It reproduces within the worker and drone cells and affects both brood and adult bees[44]. Mites spend most of their lifecycle reproducing in brood cells. It was first detected in Britain in 1992 and Ireland in 1997. Now widespread, the mite is a threat and harms the bees by feeding mainly on the fat body tissue[164], as well as acting as a vector and transmitting viruses[123,171].

### Signs/Diagnosis

Female mites are visible to the naked eye. They are reddish-brown, crab-shaped and are 1.1 mm long and 1.7 mm wide (**Fig. 4.6a**). In colonies that have not developed tolerance[129], high levels of mite infestation can cause clinical signs at both the colony and individual bee level if *Deformed Wing Virus* is present. The signs generally become apparent, in the following order, as the mite levels increase:

- increase in natural mite drop, and mites on brood removed by fork
- poor brood pattern, a sign that needs to be distinguished from other brood conditions **(Fig. 4.6b)**
- deformed bees[198] **(Fig. 4.6b)**
- mites visible on adult bees with brood present, phoretic stage
- overall decrease in bee population in a colony[115]
- parasitic mite syndrome[188]
- colony collapse

### Virulence/Spread

There has been evidence for a long time of genetic resistance of honeybees to *Varroa* mites[36,106,169]. In colonies that are no longer being treated for *Varroa* mites, tolerance will be selected for both mites and *Deformed Wing Virus* (para. 5.1)

- *Varroa* mites are widespread in the beekeeping world
- when *Varroa* first arrived as an exotic mite the initial infestation would typically kill colonies within a few years if left untreated[35], but would kill much sooner if a colony is adjacent to *Varroa*-infested colonies that are collapsing[81]

- the mite is spread naturally through robbing and drifting[81]
- beekeepers also spread the mite through the movement of infested brood or adult bees

## Treatment

There is increasing evidence that tolerance to the *Varroa* mite is being selected where there is a concerted effort within a beekeeping community to stop treating and raise replacement colonies within the area[129]. However, where beekeepers wish to treat their bees it must be in accordance with current *National Disease Policy* (para. 4.17) and manufacturer's instructions for the chemicals.

- screened floor can be used to enable the natural mite-drop to be monitored (**Fig. 4.6c**)
- mite-drop, taken over, say, three to five days, can be expressed as number of mites per day (mpd) and this is an indication of the seasonal mite populations[76]. To avoid re-infestation, all colonies in an apiary should be treated at the same time[95]
- other diseases should be kept under control[171]
- acaricides with synthetic chemicals were generally applied when *Varroa* mites first arrived in a country. Proprietary products such as Bayvarol® were widely used. Resistance of the mite to these treatments usually occurs within ten years of use[177]
- on the other hand, essential oils (e.g. thymol) and organic acids (e.g. oxalic acid) that have a different mode of action on the mite[43], can give effective control without any known resistance to date
- beekeepers using the thymol based Apiguard® in temperate climates (**Fig. 4.6d**) should be aware that efficacy is temperature sensitive and treatment should be started in early autumn (preferably during August). With a nucleus, use a half measure, that is 25g of Apiguard® gel in the tray
- oxalic acid-based products, such as Api-bioxal, should only be used when there is no brood present in the hive. They should be used as a follow-up winter treatment, if required, to the autumn treatment

**Figure 4.6a.** Photograph of *Varroa* mites at different stages on larva.

**Figure 4.6b.** Colony with high *Varroa* infestation, poor brood pattern and deformed bees.

Figure 4.6c. A squared insert taken from below a *Varroa* floor screen can collect the fallen mites, permitting a mite count to be made. Inset, close up of mites.

Figure 4.6d. Applying Apiguard®, with an eke being used to give extra height over the brood frames before replacing the crown board

## 4.7 Tropilaelaps mites *(Tropilaelaps* spp.)

### Cause

*Tropilaelaps* mites are parasites of honeybee brood. There are three species, *T. clareae*[55], *T. koenigerum*[56] and *T. mercedesae*[80], and all appear to have had the giant Asian honeybee, *Apis dorsata*, as their primary natural host before switching to European honeybee, *Apis mellifera*. To date they have not been found in Europe and are therefore *quarantine* pests; pests of potential economic importance that are being officially controlled. In their natural tropical/sub-tropical range, they have caused large economic losses on *Apis mellifera*[220]. The mites are much smaller than *Varroa* mites (**Fig. 4.7a**) but are visible to the naked eye.

### Signs/Diagnosis

*Tropilaelaps* have much in common with *Varroa* mites, although unlike *Varroa* they cannot feed on adult bees.

- mites are elongated (0.7-1.0 mm long and 0.6 mm wide), light brown in colour and the two front legs are held up like antennae[8]. They are the same length as a *Varroa* mite and about one-quarter the width and are readily identified under a x10 microscope **(Fig. 4.7a)**.
- they move fast along brood comb and tend to hide in cells rather than on adult bees like *Varroa*
- infestation leads to abnormal brood development, with adult bees having deformed wings and legs
- eventually brood and bees die, and the colony can go into rapid decline accompanied by a smell of decay in the hive

The techniques to detect *Varroa* mites can be applied to *Tropilaelaps*:

- examination of floor debris
- removal of brood with fork to check for mites
- short-term use of *Varroa* acaricides and testing for mite drop

Samples of suspect mites should be sent for laboratory examination. The samples should be killed by putting them into a freezer and then placed in a secure cardboard box along with full background details.

### Virulence/Spread

*Tropilaelaps* mites can only feed on bee brood, and hence their virulence is thwarted in situations where only adult bees are available[221]. However, where brood is present all year round the mite is highly virulent in *Apis mellifera* and can transmit viruses such as DWV[80]. In situations where it is coincident with the *Varroa* mite it will out-reproduce the latter and may thus be considered a more serious parasite.

There is no reason to assume that the mite could not exist in some parts of Europe.

- the first concern is to ensure that it is not introduced. EU legislation presently prohibits the importation of bees from Third Countries (except New Zealand)
- once in the country, its spread would be similar to that of *Varroa* mites
- in the event of the *Tropilaelaps* mite arriving, it is likely to be the subject of a specific national directive

## Notification/Treatment

*Tropilaelaps* mites must be notified to the national authority. Beekeepers should be aware of current *National Disease Policy* (para. 4.17) and apply all chemicals in accordance with manufacturer's instructions.

**Figure 4.7a** *Tropilaelaps* mite on *Apis dorsata* larva. (bar = 1 mm added)

## **4.8** Small hive beetles *(Aethina tumida)*

### Cause

The small hive beetles, *Aethina tumida* Murray[147], live in honeybee colonies and consume honey, pollen and brood[72]. The beetles are native to southern Africa where they are considered a minor honeybee pest. However, they are harmful parasites of the European honeybee and are a major threat to beekeeping. The beetles were introduced to the US in 1996, Canada and Australia in 2002. In 2014 the beetles were first detected in southern Italy. To date they have not been found in other parts of Europe and are therefore *quarantine* pests. The beetles, while much smaller than the honeybee, are large and readily visible to the naked eye.

### Signs/Diagnosis

*Aethina tumida* is a large honeybee parasite and scavenger.
- beetles are on average ~6 mm long and ~3 mm wide with a large variation in adult size[68]. They are dark brown to black in colour with young adult beetles being lighter, with club-shaped antennae and short wing cases **(Fig. 4.8a)**
- eggs are laid in clusters in crevices or brood comb and are pearly white, like honeybee eggs but smaller[70]
- larvae are creamy white, have three pairs of legs, possess rows of dorsal spines and are ~10 mm when fully grown[121]. Larvae can be found burrowing through comb or debris in large numbers and can cause considerable damage
- often smear trails are left both inside and outside the hive. There can be a rotten odour associated with the larvae.
- larvae leave the hive in masses to undertake their pupation stage in the soil[121]
- corrugated diagnostic strips can be used to aid the detection of adult mites[179]

Samples of suspect beetles should be sent for laboratory examination. The samples should be killed by putting them into a freezer and then placed in a secure cardboard box along with full background details.

### Virulence/Spread

The small hive beetles can reproduce rapidly under the right conditions. Temperature and soil humidity can be limiting factors in the pupation stage[66]. Strong colonies can remove larvae but have difficulty dealing with adult beetles within the hive, although there is some evidence that the bees do encapsulate small hive beetles in propolis[151]. The adult beetles are able to fly over several kilometres and hence a rapid spread over vast distances is possible[197].

There is no reason to assume that the beetle mite could not exist in some parts of Europe.

- the first concern is to ensure that it is not introduced. EU legislation presently prohibits the importation of bees from Third Countries (except New Zealand).
- in the event of the small hive beetle arriving, it is likely to be the subject of a specific national directive.

## Notification/Treatment

Small hive beetles must be notified to the national authority. Beekeepers should be aware of current *National Disease Policy* (para. 4.17) and apply all chemicals in accordance with manufacturer's instructions.

**Figure 4.8a.** Adult small hive beetles shown on comb along with honeybees. Inset, enlarged small hive beetle.

## Viruses

## 4.9 Chronic bee paralysis virus (CBPV)

### Cause

Aristotle reported paralysis-like signs in honeybees over two thousand years ago. However, Chronic bee paralysis virus was only identified and named in 1963 as a cause of paralysis[25]. The virus has a unique morphometry and has not currently been assigned to a particular virus family[156]. The virus has spread throughout the beekeeping world[67,17,19]. It infects the brood (eggs and larvae) as well as the adult bees, but the virus load is small at the brood level and is unlikely to have much influence on brood losses[30].

### Signs/Diagnosis.

The occurrence of CBPV tends to be sporadic. Two distinct symptoms or types are present[22]:

Type 1

- abnormal trembling of the wings and body
- crawling on the ground and on plant stems, often in large numbers
- huddling together on top of the hive cluster
- bloated abdomens and partially dislocated wings

Severely infected bees can suddenly collapse at the end of the summer leaving a few bees with the queen on neglected brood.

Type 2

- bees hairless, becoming black and with broad abdomens (**Fig.4.9a**).
- shiny, with a greasy look and nibbled by the other bees
- becoming flightless, trembling and dying in a short time

Type 1 is sometimes confused with tracheal mite (acarine) infestation. However, with tracheal mites i) the bees try to shiver their wings rather than tremble all over, ii) in temperate climates the colony dies out in late winter/early spring and not at the height of the summer and iii) the affected bees live for long periods after infestation and the crawling bees can live for many weeks when put into an incubator and fed[140,134], whereas bees with CBPV die within a few days[25].

Although CBPV has usually clearly defined symptoms, confirmation can be made using e.g. a nucleic acid (RNA) method. Samples for testing should be taken from the older bees at the hive entrance[30]. CBPV was claimed in 1964, and still disputed by some to this day, as the cause of the condition known as Isle of Wight disease[15] (para. 4.5). However, the US has had tracheal mites (IOW disease) since 1984, yet CBPV was only detected for the first time in 2010[205].

## Virulence/Spread

In the past the virus was not considered to inflict widespread colony losses but can cause severe localised deaths[167]. However, there have been reports of a recent re-emergence of the virus in many countries. In England and Wales, the number of cases rose exponentially between 2007 and 2017[37] with twice the incidence among 'professional' vs 'amateur' beekeepers. Also, those who imported queens had relatively greater increased viral incidence. In common with most viruses in honeybees, CBPV can exist at low and undetected levels for many generations[7,41,16]. However, in certain conditions (as described below) such viruses may be activated and initiate fatal infections[166]. In acute infections, CBPV can kill bees in a few days[25]. No association has been established between *Varroa destructor* infestation and outbreaks of the virus[26,201], and if it does exist, it is likely to be small[39]. Neither is there any evidence that tracheal mites transmit CBPV[23]. The virus is considered to be spread by:

- contact between bees in the colony[166] and between colonies by robbing
- conditions that cause long-term confinement in the hive[23]
- food exchange between bees[16] and
- queen to offspring (vertical transmission[27])

However, horizontal transmission (bee to bee, same generation) is considered the predominant mode of infection[30]. As with other viruses, infection through the bee's cuticle is much more potent than via the food channel. The virus has recently been detected in ant populations, and this may be a mode of infection for honeybees in some parts of the world[39]

## Treatment

Beekeepers should be aware of current *National Disease Policy* (para. 4.17). There are currently no antiviral treatments available for honeybees. Viruses persist at low levels in normal healthy colonies and can become endemic during periods of stress. Managing bees so that they can develop into strong colonies should be the intention. The situation can be improved by:

- having apiaries in good foraging locations and limiting the number of colonies per site[23]
- minimising colony disturbance (manipulation/migration)
- reducing robbing

■ keeping other diseases under control

Neither *Varroa* mites[26,201] nor tracheal mites[23] have been shown to be vectors (carriers) of CBPV.

Further improvements can be made by:

■ replacing frames on, say, a 3-year cycle (all together rather than individually)
■ re-queening for improved resistance[16,117,168]

**Figure 4.9a.** Black, hairless bee in the centre of the photograph showing signs of CBPV (Type 2).

## 4.10 Sacbrood virus (SBV)

### Cause

Sacbrood was first identified in 1917 in European honeybees[216]. A strain of the virus is also present in the Eastern honeybee *Apis cerana.* It is caused by Sacbrood virus (SBV) which when fed to larvae can prevent them from pupating and results in death.

### Signs/Diagnosis

The condition has clearly defined clinical signs[188]:

- only sealed brood are affected
- cappings can be perforated where the bees attempt to uncap the cells
- diseased larvae change from the normal pearly white to pale yellow-brown fluid sacs and heads darken **(Fig. 4.10a)**
- larvae dry to a "slipper-like" dark-brown scale

### Virulence/Spread

The virus is not considered virulent at the colony level. It is spread by contaminated adult bees feeding young larvae. The bees normally detect the diseased larvae and remove them at an early stage. In addition, the virus appears to lose its infectivity quickly in the dry state[22].

The virus is currently widespread in the beekeeping world. There is evidence that the incidence of SBV is increasing in some European countries[1152].

### Treatment

Infected colonies can recover without treatment. Badly infected colonies should have the brood comb removed and replaced with foundation. Re-queening the colony should be considered where a lot of comb has been seriously affected.

Figure 4.10a Heads of SBV infected larvae sticking out of the brood cell.

## 4.11 Deformed wing virus (DWV)

### Cause

Deformed wing virus (DWV) has only been isolated and named in the early 1980s[22]. To date four different strains have been identified; DWV-A, DWV-B (formerly *Varroa destructor* virus-1 or VDV-1), DWV-C and DWV-D (Egypt bee virus)[29,57]. DWV infects all the castes: workers, drones and queen, and has a whole-body effect on the honeybee. It has been detected in bumblebees[89], *A. cerana* and the dwarf honeybee, *A. florae*[7], as well as *V. destructor*[31], *Tropilaelaps mercedesae*[50] and the small hive beetle[73]. The virus has had little affect on the honeybee prior to the arrival of the *Varroa* mite, which has acted as a vector. It is now the most globally abundant virus[104].

### Signs/Diagnosis

There were no clinical signs of this virus before the arrival of *Varroa* mites[57]. Signs of *Varroa*-mite vectoring became apparent shortly after *Varroa* became established[7]. Signs include:

At individual bee level:
- death of pupae in the cells or adult bees within 2-3 days of emergence
- deformed/shrivelled wings and other appendages **(Fig. 4.11a)**
- shortened and bloated abdomens
- discolouration in adult.

At colony level:
- in acute cases[222], the *Varroa* mite plays an important role in the "parasitic mite syndrome" collapse of the colony
- masses of dead or dying bees
- pupae failing to emerge
- spotty or neglected brood

At high levels of infection, DWV has usually clearly defined symptoms; however, confirmation at low levels can be made with diagnostic techniques such as the nucleic acid (RNA) method. Samples for testing should be taken from the older bees at the hive entrance[30].

### Virulence/Spread

DWV has spread throughout the beekeeping world. In the absence of *Varroa* mites, it is generally of low virulence. The virus is spread by:
- vertical transmission, through drones and queens[223]

- horizontal transmission, between bees, largely through food transfer

When vectored by *Varroa* mites it is spread by:

- horizontal transmission (bee to bee), but the mites permit a much more virulent form of the virus to establish that can kill colonies[7]

Originally, wing damage was considered to be due to *Varroa* mites feeding during pupal development[3], but it is now accepted that damage is due to DWV particles being introduced through the cuticle during mite feeding[7]. The dominance of *Varroa destructor* in the spread of DWV is shown by the fact that the seasonal distribution of the virus follows closely that of the mite[153,201].

## Treatment

In the absence of *Varroa* mites, the virus is not a serious threat to honeybees. DWV will persist in bees even after the mites have been removed, hence, where treatment is still being used, it should be undertaken in early autumn[198]

**Figure 4.11a** Photograph of emerged bee with deformed wings in a heavily varroa-infested colony.

## 4.12 Other viruses

Some eighteen viruses[40] had been identified by 2007, largely those revealing themselves through signs/symptoms: physical, developmental, behavioural or demographic. Since then, using cheap and fast, high-throughput sequencing (HTS) technologies many more viruses have been identified[29]. The overwhelming majority of this wide diversity of viruses is asymptomatic and only a small proportion causes disease. Those that have been identified as causing disease in the honeybee, include the three listed above, DWV, SBV and CBPV. A further six of the more important viruses are considered here.

Most viruses were only of scientific interest until the arrival of *Varroa* mites introduced a new transmission route. Evidence of the direct relationship between DWV and *Varroa* mites is readily apparent to the observant beekeeper. A group of three closely related viruses that largely depend on *Varroa* for transmission are **Acute bee paralysis virus (ABPV), Kashmir bee virus (KBV)** and **Israeli acute paralysis virus (IAPV).** These viruses are globally widespread and have been implicated in colony deaths, especially in conjunction with high levels of *Varroa* mites. More recently, IAPV has been claimed to be a specific indicator of colony collapse disorder (CCD) in the US[46].

While globally widespread, these viruses have generally been at a low prevalence with no obvious symptoms at either individual or colony level. However, they can be extremely virulent in conjunction with *Varroa* infestations. Rapid adult mortality is common for all three viruses, but normally there is no evidence at the larval stages[58]. With ABPV and IAPV, but not KBV, death is preceded by a progressive paralysis, including trembling, inability to fly and hair loss on the thorax and abdomen. There is seldom mass paralysis at colony level as with CBPV, presumably due to the relatively rapid progression from paralysis to death. Diagnosis can generally only be confirmed with molecular protocols, such as reverse transcription, polymerase chain reaction (RT-PCR).

As these viruses are only expressed in association with *Varroa* mites, having *Varroa* tolerant bees[129] or keeping the mite levels under control provides effective remedy. Research is ongoing to develop a possible treatment using technologies, such as ribonucleic acid interference (RNAi). However, these technologies are expected to have associated difficulties.

**Slow bee paralysis virus (SBPV)** is also associated with *Varroa* infestation and infects both larvae and adult bees. However, once well established, it continues to multiply within the colony even after mite numbers are reduced following treatment, and the colony can die out later in the year. Dead shrivelled-up larvae will be in sealed cells and infected adult bees will show paralysis and can die very quickly.

**Cloudy wing paralysis virus (CWV)** has been associated with collapsing colonies infested with *Varroa* mites. However, there is no definitive evidence that mites are a vector in its transmission[38]. Infected bees can have opaque wings, but this is not a reliable diagnosis. While it has been reported from many countries throughout the world, its prevalence is currently relatively low.

**Black queen cell virus (BQCV)** has long been associated with the microsporidian *Nosema apis*[22]. The queen cells develop dark brown to black cell walls, and at the early stages the dead pupae resemble sacbrood. Adult workers, drones and the queen are infected, and transmission is both horizontally (mainly through feeding) and vertically (queen to offspring). The close relationship between seasonal variation of *N. apis* in colonies and the incidence of BQCV is striking, with summer peaking being the norm. It is controlled by keeping *N. apis* in check (para. 4.4). There is no evidence that *Varroa* mites are a transmission route.

## Other disorders

## 4.13 General

### (a) Amoeba (*Malpighamoeba mellificae*)

Amoeba is caused by the protozoan *Malpighamoeba mellificae*. The parasite enters the gut of the adult bee in the form of cysts, germinates and passes to the malpighian tubules[22]. It is similar in its spread and behaviour to *Nosema* spp. but its prevalence in colonies nowadays is considered low, and it does not appear to have any appreciable affect on the bees.

### (b) Parasitic mite syndrome

This condition was first identified in the US at the Beltsville Research Laboratory[188] in the 1990s after the two parasitic mites, tracheal and *Varroa*, had devastated honeybee colonies. There is an array of clinical signs such as a reduction in colony population and queen supersedure, but it is usually most apparent in the spotty brood pattern. The larvae can go through stages akin to an infection of foul brood. (A personal experience of this condition was when colonies were going through a phase of tracheal-mite infestation that coincided with the first appearance of *Varroa* mites in the apiary. A large proportion of brood had melted down like an AFB attack but was not ropey.) If noticed in time colonies will revert to normal when the mites have been brought under control.

## (c) Stone, bald, chilled brood

The beekeeper should become familiar with as many brood conditions as possible in order to identify cause and take remedial action.

**Stone brood** is a rare occurrence in honeybees. It is usually caused by either of the fungi *Aspergillus flavus* or *A. fumigatus,* which occur commonly in soil and cereals[22]. The fungi affect bees and can cause a respiratory disease in humans and other animals if the spores are inhaled. The condition is similar in many ways to chalkbrood (para. 4.3) but the former causes larvae to turn into very hard mummies.

**Bald brood** is normally caused by wax moth larvae as they travel over sealed comb leaving a trail of uncapped brood. Bee larvae usually complete their pupation and occasionally emerge with deformed wings and legs[22]. Controlling the incidence of wax moths (para. 4.13e) in the colony will remove the problem.

**Chilled brood,** as the name implies, is caused by brood of all stages dying after being exposed to a period of low brood temperature. Normal brood temperature is 34.5 ± 0.5°C[218]. However, brood from a predominately *Apis mellifera mellifera* subspecies, during an experiment, did pupate and emerge in an incubator with no clinical signs while being held as low as 30°C[135,140]. These adult bees will not be viable[200] and will have a reduced lifespan. Hence, temperatures below 30°C are required to cause chilled brood and the condition is therefore likely to be as a result of brood at the periphery of the brood nest being exposed to low temperatures. This can be due to a sudden drop in bee population (from disease or pesticides) or a sudden drop in ambient temperature particularly in early spring. The dead brood can be at all stages of development and the uncapped brood can turn very dark in colour.

An extreme case of chilled brood is where there is, in effect, a chilled colony situation and this is a common cause of colonies dying out. Where a reduced bee population in a colony is combined with low ambient temperatures such as in early spring, this can give rise to a situation where the bees become isolated and are unable to move to stored food on the comb and perish. Brood and bees can be affected, and it may involve the whole or part of the colony (**Fig. 4.13a**). This is often referred to as isolation starvation.

**Figure 4.13a.** A chilled colony where all or part of the colony is isolated from the food source and starves.

**Figure 4.13b.** A honeybee on a landing board in early spring showing a trail of fresh dysentery as well as old, dry dysentery.

## (d) Dysentery

The dysentery condition in bees is present when they leave excrement on the outside of the hive **(Fig. 4.13b)** or in more severe cases on the inside hive parts. There is no need for any parasite or pathogen to be involved. It results from an excess of water in the intestine and causes may include unripe or granulated stores. It may also be from storing a lot of waste matter in the rectum during a long winter confinement. It can be associated with parasites such as tracheal mites or *Nosema apis*, but there is evidence that it is uncommon with *Nosema ceranae*[103].

## (e) Pests; wax moths, mice, wasps/hornets

Pests usually take advantage of weak colonies, and in conjunction with other conditions or parasites that may be present can have a cumulative and synergistic affect.

## Wax moth

There are two wax moths that infest honey colonies. The greater wax moth, *Galleria mellonella* (Insecta: Lepidoptera), and the lesser wax moth **(Fig. 4.13c)**, *Achroia grisella* (Insecta: Lepidoptera). The wingspan of the greater is about 30 mm and the lesser about 20 mm. In the evolution of the honeybee, wax moths provided a valuable service in removing old-infected combs from colonies that had died out **(Fig. 4.13d)**.

The larvae of both moths will burrow through wax comb leaving a trail of their faeces behind. The larvae of the greater wax moth will also attack the hive woodwork and excavate hollow trails. In addition, the moths will also tunnel across sealed brood and cause the condition known as bald brood (para. 4.13c).

The moths can be controlled in the hive through the beekeeper having strong colonies and building up nuclei quickly. If stored comb is stacked outside, then cold weather conditions will usually keep the moth under control. Acetic acid fumes will kill most stages of the moths' development. Moths generally do not attack wet comb so "wet" supers, which are not returned to the colony to clean up, are normally not affected. Sealing supers in plastic bags immediately after extraction or after returning "dry" from the apiary will usually remove any moth problems **(Fig. 6.2d)**. Freezing the combs at -20°C for up to a day will kill all stages of the wax moth. Putting the comb into sealed plastic bags before freezing and allowing it to return to room temperature before removal from the bag will reduce the likelihood of condensation on the comb.

## Mice

Mice and rat damage is generally a localized problem. It is, however, prudent in many areas to put on a mouse guard in late summer/early autumn before rodents migrate from outdoors. A metal guard will normally provide an effective barrier against large rodents **(Fig. 4.13e).**

## Wasps / hornets

The common wasp (*Vespula vulgaris*), can attack weak or queen-less honeybee colonies, rob the honey and even overwhelm the colonies. Their populations build up during the year and are at their peak towards the end of the summer. It is widespread throughout Britain and Ireland, while the German wasp *Vespula germanica* is also present. (Of course, other honeybees rob colonies, and bumblebees (*Bombus* spp.) also rob but to a lesser extent). The prevention and control of robbing is discussed in paragraph 5.6.

The European hornet, *Vespa crabro,* is generally considered to be much less aggressive than the common wasp. It has been in Britain for over ten years, is widespread in the south/central area and expanding northwards. It is not known to be in Ireland.

The Asian hornet, *Vespa velutina*, is spreading across mainland Europe, since first accidentally being introduced to France in 2004. While it has been sighted since 2016 in Britain it appears not to have been fully established by early 2021. However, in early May 2021 an insect identified as an Asian hornet was found 'alive but dying' in a north Dublin suburb. This calls for ongoing vigilance by beekeepers and the public at large to thwart the hornet from becoming established. (Identified as book was going to printers). It comes with a reputation for aggression and in time there is no reason why it should not continue its colonization of Europe **(Fig. 4.13f).**

Recently (2019), the Asian giant hornet (*Vespa mandarinia*), a social species of the same genus as *V. velutina*, has been discovered in the north west coast of America[5]. It is the largest hornet in the world and preys on honeybees, and is predicted to have a serious effect on US beekeeping.

## (f) Chemical poisoning

Poisoning may be due to natural substances such as toxic nectar, but it is generally caused by exposure to spray or dusting of fungicides, pesticides or herbicides. Honeybee deaths usually result from the chemicals affecting the nervous system and causing a breakdown in the bees' co-ordination which can be confused with paralysis. The alimentary system is normally affected, leading to starvation. The symptoms of poisoning are typically:

- a large number of dead or dying bees suddenly appearing at the hive entrance
- bees being ejected from inside or refused entry
- dead bees with proboscises extended, and
- crawling, trembling, confused bees outside the hive. In the case of natural poisoning, the signs are not usually as pronounced, and bees can be found near the offending plants. Death due to disease usually involves bees dying over several days, with bees outside the hive in varying degrees of decay.

When poisoning is suspected, the beekeeper should record as much detail as possible, take photographs, sample pollen, estimate the number of bees affected etc. Collect a sample of at least 200 dead bees. Taking a sample of bees is discussed in paragraph 5.7. In the UK there is a comprehensive reporting system under The Wildlife Incident Investigation Scheme (WIIS) www.defra.gov. uk/fera

Poisoning incidents vary widely in their impact on the colony. In severe cases, where both foraging and house bees are affected, the colony is unlikely to survive. If forages only are involved, the colony will usually recover in a few weeks but may require immediate feeding.

All steps should be taken to avoid poisoning in the first place. The beekeeper should be alert to what is happening in the neighbourhood of the apiary and the crops that are in the immediate area. It is critical to liaise with farmers and horticulturists regarding crop-treatment plans and generally establish a good local network. Crop owners will normally be co-operative and will frequently agree to spray crops outside normal bee flying times or at least give good advance notice of any spraying activity.

**Figure 4.13c.** A lesser wax moth (*Achroia grisella*) among debris on the hive floor.

**Figure 4.13d.** A frame of wax comb after an infestation by wax moths

**Figure 4.13e.** A metal mouse-guard fitted to the hive

**Figure 4.13f.** Asian hornet (*Vespa velutina*) is an aggressive member of the hornet family and has been in Europe since 2004.

## 4.14 Queen conditions

### (a) Queenlessness

When there is no laying queen in the hive, there will be an immediate change in the brood pattern, with possibly no eggs and diminishing brood. The beginner-beekeeper can become very impatient and may look for a new queen to introduce to the hive. Establishing that there is no queen in the hive can be a difficult task. The colony may display restlessness, be more aggressive than normal and not be making preparation for a laying queen by polishing patches of cells. There may be an abundance of cells in the brood box filled with pollen and nectar.

The tried and tested method of detecting queenlessness is to introduce a frame of eggs and young larvae (taken from another colony with workers shaken off) into the centre of the brood nest. For most of the year, this can give a fairly reliable indication. If queen cells are being drawn out on this "test frame" within a few days, queenlessness is indicated. The small-scale beekeeper, having limited resources, will frequently allow a queen to be raised from a selected queen cell on this frame, whereas those operating on a larger scale may knock down all queen cells and introduce a laying queen.

If no queen cells are drawn out on the test frame, it may be that a young queen has emerged after a colony has swarmed and is not yet laying. Introducing a test frame in situations where no eggs are present up to four weeks after the young queen has emerged will usually indicate the status (recently the young queens appear to be taking longer to get into lay, and beginner-beekeepers need to exercise patience, subject to their being no signs of laying workers). If no queen cells are drawn, it may mean that a queen has been present for a long time but is not laying. The queen should be found, removed and a queen cell or laying queen introduced.

### (b) Drone-laying queen.

It can happen either through age or infirmity that a queen will be unable to fertilise her eggs and drone brood will result. Thus, although a queen is present, drones will emerge from worker cells. Hence, a brood pattern that may start with only a few drones among worker cells **(Fig. 4.14a)** will become increasingly drone-dominated and may ultimately be all drone. The queen needs to be found and the colony re-queened.

**Figure 4.14a.** Drone laying queen showing a lot of drone capping (domed) on worker cells along with worker capping (flat) on worker cells

**Figure 4.14b.** Laying workers showing several eggs in a single cell with some laid on the cell walls. Inset, close-up of cell with two eggs

## (c) Laying workers

Where a colony has no queen and does not have the potential to raise a queen, some workers in the colony may start laying. The eggs will be unfertilized and will therefore only produce drones. The condition will be apparent in the early stages by the appearance of eggs (sometimes multiple eggs) mostly laid on the side of the cell (**Fig. 4.14b**). Later raised drone cappings will appear on worker cells and eventually a patchy laying pattern on several brood frames. If no action is taken, the colony will die out and a lot of brood comb will be ruined.

An established laying-worker colony is very difficult to re-queen in the normal manner. Prevention of the condition in the first instance should be the approach. In a situation where a new queen is being raised by the colony and there are no eggs laid within three weeks of the young queen emerging, a frame of eggs and brood should be given to the colony to allow it to bring on a queen (see Queenlessness above).

Where laying workers are present, a number of approaches have been used in the past. In the early part of the season, combining the colony by placing the hive on top of the supers in a strong colony can be undertaken, but this would only be recommended at the early stages of laying workers. A traditional approach has been to shake out the colony at some distance from the hive location and re-queen the bees that return. From personal experience this approach is a waste of time and maybe of a new queen.

An approach that has given some success is to shake the bees off all the frames into the hive. Then flush out all the brood from the combs with a strong jet of water and shake out the frames. Undertake this operation outside the hive, or off-site if necessary, and this will remove the old brood pheromone. Return the frames to the hive and spray the bees with water on each side of all the combs and insert one or two frames with eggs and brood from a healthy colony and let them raise their own queen. Alternatively, introduce a queen cell (there can be success introducing a new queen in a cage) but this is only likely to be successful at early stages of queenlessness. Feed if there are no supers on the hive or the colony is in a nucleus hive. There will be a lot of water around but out of the confusion a satisfactory re-queening should occur. Again, only undertake this exercise at the early stages of laying workers, as a dwindling colony with messed-up brood comb is not worth the effort.

## 4.15 Colony collapse disorder

Colony collapse disorder (CCD) is a term used to describe a phenomenon that was observed among honeybee colonies in North America[46]. In the winter of 2006-2007, large-scale losses of managed colonies were reported, and these losses continued over the following winters.

There was a common set of symptoms[208]:

- rapid loss of adult bees, leaving weak or dead colonies with a relatively large amount of brood
- lack of dead worker bees around or in the hives
- delayed invasion by hive pests such as wax moths

Many claims have been made as to the cause including: *Varroa* mites, pesticides, viruses (mostly Israeli acute paralysis and invertebrate iridescent viruses), *Nosema* (mostly *Nosema ceranae*)[159], environmental factors and beekeeping practices.

After almost 5 years of research, there is still no consensus as to the cause. It is uniquely a North American problem, despite occasional claims from other parts of the world, including Europe. The search for the cause has been confounded by the many variables involved, not least the large number of parasites now associated with honeybees[46]. There is an emerging view that CCD involves an interaction between honeybee parasites and other stress factors, including "modern" chemical insecticide and herbicide treatments used in both agriculture and apiculture[8].

## 4.16 Disease transmission

Bee diseases do not crop up spontaneously from within a colony. They are transmitted to the colony by bees or beekeepers. Here are some personal experiences to illustrate the different modes of transmission.

In September 2008 the author helped a beekeeper to take samples of bees for disease testing. The report came back positive for *Nosema ceranae*, the first time that it had been found on the island of Ireland. Subsequent telephone calls were received from known importers of Buckfast bees from around Ireland to enquire if the new strain of *Nosema* had been found. A beekeeper adjacent to the beekeeper with the positive test had been an importer of Buckfast queens for decades!

*Varroa* mites were first detected in Ireland in June 1998 in County Sligo and in County Mayo in the following spring. Importation of an infested colony was suspected. The author's out-apiary is in County Galway and it was there in September 2000 that the first mite in County Galway was detected. On the other side of the country, in County Carlow, a fresh outbreak was detected at the end of 1999 and was thought to have been there for some time. Again, importation of an infested colony was suspected. The author's home apiary is north west of Swords in County Dublin and the first detection of the mite, north of the Liffey, occurred there in September 2003. It was later discovered that a few infested colonies had been moved from County Wicklow to within 300 metres of the apiary. No mites had been detected on the floor debris in mid-summer. The first sign that something was wrong was the presence of a screen, a curtain of propolis plastered on the inside front wall of one of the hives leaving only a small entrance. This was the bees' response to the trauma of an invasion of mite laden bees, probably absconding from one of the nearby hives. The Bayvarol mite drop was over six thousand in a ten-day period. As a result of the infestation, the author lost six months (had to repeat) of an experiment researching small-cell combs.

American foul brood (AFB) is arguably the most virulent of our bee diseases. One weekend towards the end of January a telephone call was received from a beekeeper that the author had been mentoring to say that his hive had increased in weight by almost four pounds since the previous weekend. He had used a direct method of weighing where the stand mechanism lifted the hive fully off the stand, so the weight increase had to be taken seriously. The previous week had been unseasonably benign, no winds, very sunny and temperatures up to 12°C. But it was still winter and there were no forage plants of any significance. At the beginning of April, another call was received from the same beekeeper to say that he had detected AFB in his hive. A photograph confirmed a large patch of darkish sealed brood on one comb (a bit reminiscent of early blight on a potato tuber) that tends to depict robbed honey from an infected colony. After an AFB alert went out for members to check their colonies, it was reported that a nearby colony had died out sometime during the wintertime and had not been sealed up. The original infected colony had been imported as a nucleus, as had all the outbreaks of AFB over the past 20 years in the area. In two cases they had been nuclei, the first colonies for beginners, purchased from 'reputable' suppliers down the country. In each case AFB had been detected within three months of purchase.

A range of transmission modes are depicted above, from national transmission by robbing and movement of colonies, to international transmission through the importation of queens and colonies. Responding to these, the local association has for over fifteen years been discouraging the importation of colonies into its catchment area (para. 5.1). It also illustrates the need for collective responsibility among beekeepers as we are dependent on each other for our bee health.

## 4.17 National disease policy

Beekeeping is governed in most countries by national legislation. In the EU there are three separate strands of legislation: regulations covering food and food hygiene (para. 6.2), animal remedies (para. 5.8) and diseases (Part 4). The disease legislation lays down how diseases are to be managed and may include details on notification and treatment, including use of chemicals or destruction of colonies, as well as compensation for losses in some cases. Beekeepers should be familiar with their *National Disease Policy*, keep up to date with current policy and be aware of the national body that has been charged with its implementation.

National policy relating to diseases can change over time due to the arrival of new diseases or new treatment methods, usually relating to the use of chemicals. The relevant website should therefore be checked on a regular basis to obtain the current national position.

For the EU, broad policy is laid down within which national governments must operate. Council Directive 92/65/EEC lists American foul brood, European foul brood, *Aethina tumida* (small hive beetle) and the *Tropilaelaps* mite as notifiable diseases for bees within the community.

The parent legislation for **Ireland** is the Bee Pest Prevention (Ireland) Act, 1908 and the implementing regulations are the Bee Pest Prevention (Ireland) Regulations, 1909. These have been updated and amended through later Statutory Instruments (SI). The Department of Agriculture, Food and Marine (DAFM) are charged with the implementation of legislation; www.agriculture. gov.ie

The governing legislation in the **United Kingdom** is the Bees Act 1980, which empowers Ministers to make Orders to control diseases and pests affecting bees and provides power of entry for authorized persons. The Bee Diseases and Pests Control (England) Order 2006 (SI 2006 No 342) empowers the Department for Environment, Food and Rural Affairs (Defra) to control AFB and EFB in England. There are similar Orders for the rest of the UK. These orders are implemented separately by Government Departments in England, Wales, Scotland and Northern Ireland. In **Northern Ireland**, it is the responsibility of the Department of Agriculture & Rural Development, NI; www.dardni.gov.uk and in **Britain**, Defra; www.defra.gov.uk

# PART 5 | Working with the bees

## 5.1 The bees & sustainable colonies

As already mentioned, *Apis mellifera mellifera* was the subspecies that originally inhabited northern Europe (para 2.1). In the last century, possibly triggered by the onslaught of tracheal mites (Isle of Wight disease) in the early 1900s, an increasing movement of honeybee queens and colonies was experienced within Europe. The indigenous *A. m. mellifera* was replaced to a greater or lesser degree by imports, many of them hybrids. However, today in Ireland and in parts of Britain there are still relatively pure populations of *A. m. mellifera*[34,101]. It makes sense to work with the indigenous bee that has been selected over time to best cope with conditions in these areas. A recent pan-European study showed that *locally adapted* bees generally performed better and particularly so in the case of survival and parasite/pathogen level[144a]. The Native Irish Honey Bee Society (NIHBS) was set up in 2012 to support and promote the various strains of the native bee throughout the island of Ireland[149]. The society has been encouraging members to have their beekeeping association's catchment area designated as a Voluntary Conservation Area where only the native bee strains would be used.

At an international and national level, the movement of colonies brings the threat of transmitting disease (para. 4.16). It is interesting to look at an approach introduced into North County Dublin. Since ~2005, primarily to contain the incidence of American foul brood (AFB), which in each case could be traced back to bees coming from outside the area, the local beekeeping association took a policy decision to discourage the movement of bees into the area. Most imports had been by beginners, who at that tender stage are keen to get started and can be vulnerable to accepting anything that *purports* to be a bee colony, irrespective of its condition. Beginners were provided with mentors and supplied with a starter nucleus from within the area (para. 3.5). The bees in the area were generally the native *A. m. mellifera*, and up to ~2005 the strain had been enhanced by the introduction of significant numbers of black Galtee queens (that had been raised by the Galtee Bee Breeding Group in County Tipperary).

With the onset of varroa, this policy also held out the potential for natural selection of colonies with tolerance to the mite. The experience with tracheal mites in Europe since the mid-1900s showed that generally with no treatment being applied, and mite-infested colonies being allowed to die out, selection of tolerant/resistant colonies occurred. The selected colonies were those with the ability to detect the presence of the tracheal mite and to groom it off[131,162]. A similar resistance to tracheal mites occurred in the Primorsky region of Russia where bees were exposed to the mite since 1922. Also, in Russia, the Primorsky bees that were exposed to the *Varroa* mite from the early 1900s developed tolerance[154]. Studies undertaken in Sweden and France have demonstrated the ability of colonies to survive for many years without any measures taken to control *Varroa* mites[85,119]. Feral colonies have also been developing tolerance where colonies of these bees are living generally in isolated forest locations[183]. Therefore, where a large number of colonies are contained in an 'enclosed' area, where new colonies into the area are restricted, where colonies are receiving no treatment and where colonies are raising their own replacements, tolerance to the mite is selected. This is increasingly the situation in North County Dublin over the last number of years. A co-operative approach is necessary as drones are the *flux* influencing the welfare of colonies over a wide area and selecting tolerance in drones to *Varroa* mites is critical[107]. The regime described above is not simply an approach for the *Varroa* mite but will apply to all other conditions that militate against colony health and development[141a]. Colonies will also be locally adapted.

Mite drops can still be quite high, but this is the situation with the Asian honeybee, *Apis cerana*, the original carrier of the *Varroa* mite. However, there is a balance, and while there are fluctuations in the mite drop levels throughout the year the pattern is similar from year to year. During the first two years of non-treatment, dead honeybees outside the hives were to be expected and DWV was common with purges of dead bees with distorted wings mostly in the autumn. Little evidence of DWV has been observed in recent years, and a 2017 study using RT-PCR detected no DWV infection in samples of callow bees and negligible levels in flying bees[129]. Over 40% of dead natural-drop mites had legs missing or mutilated (**Fig 5.1a**). Increasingly the practice with beekeepers in the area is not to treat their bees. (Treating bees is only good for the colony, in the short term, but bad for the species) This collective effort contributes to the raising of tolerant bees, as mentioned above. The local beekeepers in effect *own* this critical resource in the drone congregation areas and contribute to its improved tolerance status over time.

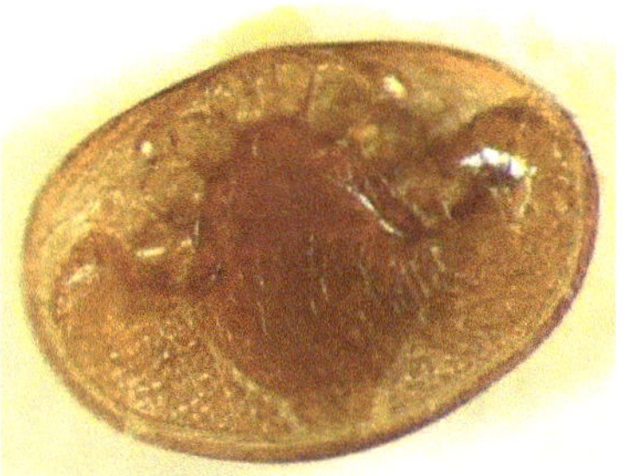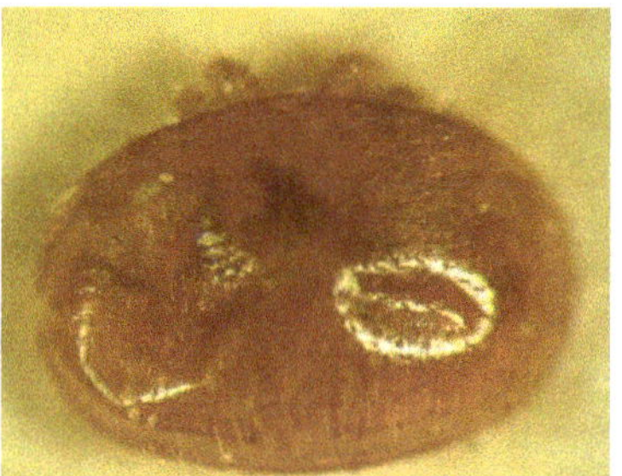

Figure 5.1a Examples of natural *Varroa* mite drops in a tolerant colony:
(L) with seven of eight legs missing, and (R) a damaged idiosoma.

The approach depends on solidarity within the beekeeping community, namely a critical mass of beekeepers in the general area not treating their bees and not bringing in non-local strains. If the situation changes, approaches such as a varroa-specific artificial swarm[76] (para. 5.2) or selecting colonies for lower mite prevalence may be necessary as colonies will be vulnerable if DWV is present. The author is now in the second decade of non-treatment with no colony selection for low mite-drop level, low swarming, honey production etc. As a general rule, each colony raises its own replacement colony (para 5.2). The lack of mass queen raising by grafting or other means has meant that genetic diversity is less likely to be diminished. The bees are making their own selection in the same way as they had been doing up to the time that skep beekeeping started. This is how the *A. m. mellifera* subspecies evolved in a temperate (cool and damp) climate and in harmony with its environment. Interestingly there is no specific selection for honey production ability. This is an innate quality that colonies evolve in any case, as those that are good foragers will survive and prosper at the expense of other strains they compete with for limited forage. Their young queens will be mated, and their drones will tend to be in the air earlier in the season and in greater numbers. But they will select for honey production traits in a balanced way, allowing for other desirable traits such as disease resistance. Colonies that have been manually selected will, in the absence of this selection, tend to revert in time to their natural niche.

Presumably with the increased purity of the strain and reduced hybridization, colonies have become more docile generally permitting bare-handed manipulation throughout the year, one of the great pleasures in beekeeping.

## 5.2 Raising colonies & swarming

There are several approaches available to beekeepers to raise new or replacement colonies. These largely depend on the scale of the beekeeping regime (para. 2.5). The approach in larger scale operations will include bulk queen raising and bought-in queens or start-up nuclei (package bees). The approach to mass queen raising is covered in other texts[105]. Using a mass production approach, along with importing bees into the area, will tend to move colonies away from being sustainable (ability to select) through a reduced and altered gene pool.

Reproduction by swarming is the natural way for honeybees to raise new colonies. During swarming about half the colony leaves the parent hive and travels to a nearby site. Exceptionally, supersedure can occur where the colony requeens without triggering the swarm instinct. However, in the absence of supersedure, due to the need for honeybees to be raised at a high temperature (~ 35°C), scale in honeybee numbers is required. Hence reproduction by swarming or division.

### Swarm prevention

It is generally accepted that swarming is initiated when the level of *queen substance* in the colony drops below a threshold value. Queen substance is a group of pheromones secreted by the queen and passed around the colony through trophallaxis, food sharing. It attracts workers and signals the queen's presence. Its intensity will be reduced in a congested colony or with an ageing queen. Giving the colony additional space by supering in good time can reduce the likelihood of swarming and can be an effective *swarm prevention* method. However, at some stage reduced queen substance will stimulate the workers to raise queen cells, mostly on existing queen cups and this can start off the swarming process. Where the *last* inspection showed a colony was *not* making swarm preparations (no egg in a queen cell), a seven-day inspection will be required during the swarming season for an unclipped queen, while this can be extended to fourteen days for a clipped queen (para. 5.3). In the latter case the queen may be lost in the swarming event, but the swarm will return to the parent hive.

### Swarm control

The presence of a charged queen cell (indicated initially by a dampness in the cell as the workers start to feed the developing larva) shows that the colony is on a swarming course. The colony will generally be ready to swarm on day eight, when the first queen cell has been sealed, but swarming will normally be postponed in poor weather conditions. A colony raising practice that can be used by small-scale beekeepers is to wait for queen cells (QCs) to be raised by the colonies and in effect to use it as a *swarm control* method. Using swarm queen cells means that the beekeeper is using cells that the colony has chosen and raised itself. Retaining the old queen, as happens in nature, requires that a seven-day inspection cycle be used. (However, this is only the case until swarm preparations have started. If queen cells have been destroyed by the beekeeper, the colony can build a queen cell on three-day old worker larvae and swarm out in two or three days) A method that is effectively re-queening "in situ" can be used as illustrated and described below **(Fig. 5.2a)**.

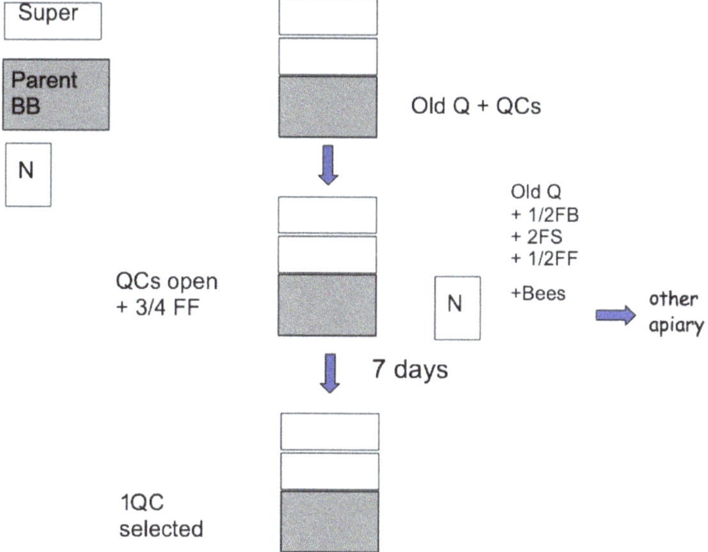


**Figure 5.2a** Re-queening "in situ", by taking out a nucleus from the parent colony. FB denotes brood frame, FS frame of stores and FF frame of foundation.

**Figure 5.2b** The two queen cells in the top left of the lower photograph had their locations marked (before being sealed) on the top bar, as shown in the upper photograph. Wax was scraped off the top bar and the queen cell locations marked by arrows depicting the side that the cells were on as well as the distance down the comb. The distance was scaled by taking the width of the top bar as equivalent to the full depth of the brood comb.

- when queen cells (QCs) are raised by the colony, go through the brood box and mark an X on top of all frames with QCs, using the hive tool. Find the queen (much easier if she is marked para. 5.3)
- remove the queen into the centre of a nucleus box with one or two frames of emerging brood (cut out any QCs on these frames) and add two frames of stores/pollen, one on each side. Shake in two frames of bees (not from those marked as having QCs) and fill out box with one or two frames of foundation or drawn comb. Nucleus box can be moved to another apiary
- alternatively, stuff entrance with grass to delay flying, move box to one side in the apiary, and check that they are flying after 24 hours or release at dusk
- cut out any sealed QCs. in the parent colony on the original location (ensure that there is at least one suitable open QC). This is a job that must be done thoroughly as even what may look like a miserable QC that a beginner may

ignore, to the colony it is a QC. Use brush to move bees to give clear sight of combs

- ■ choose an open, well-developed QC and mark its position. A useful method is to scrape the wax off the top of the frame directly above the QC and using a marker (preferably white) draw an arrow to indicate the side of the frame that the QC is on, and also scale its distance down the frame with a mark on the arrow (**Fig. 5.2b**). Other well-developed open QCs can be marked, if available, and where it is intended to raise ripe QCs

- ■ the parent colony can be made up with spare combs into the brood box and queen excluder/supers returned

- ■ seven days later the parent colony should be examined, and the marked QC(s) identified. All other unmarked QCs must be cut out. Shaking all unmarked frames to remove some bees will help in finding QCs. Check that the chosen QC has not shrunk or has signs of *Black Queen Cell virus* and since more than one QC will usually have been marked there should be options. Any additional QCs can be used to make up mating nuclei and taken away. The chosen QC can be left in the parent hive and the supers returned. In normal circumstances a virgin queen will emerge, be mated and take over the colony

- ■ egg laying by a new queen may take a lot longer than the minimum ten or so days after the virgin queen has emerged, a time lapse that appears to be increasing in recent years. Is the colony polishing up patches of cells to prepare for egg laying? Patience is required and the beekeeper should be familiar with the conditions of queenlessness, laying workers, drone layers (para. 4.14)

- ■ requeening can also be done with the established artificial swarm method[105]. A varroa-specific artificial swarm is also detailed below.

- ■ using additional QCs to make up mating nuclei, such as Apideas™ is a good way of raising additional queens, and a timely introduction for beginners in working bees barehanded (**Fig. 5.2c**). No smoke, one fine mist of water as crown cover is prized off, and a fruit punnet for parking removed comb and holding them upright.


The overall approach is simple, does not require much equipment or space and is an effective way of swarm control. There is no requirement to lift the full parent hive to one side as in the classic artificial method. As the parent hive is not moved this is a re-queening *in-situ* method. Honey production is usually not greatly affected. The nucleus box with the old queen can be used to make increases or made available to other beekeepers or beginners, as it is a colony of known temperament, important for someone starting out. The approach mimics that of a feral colony with no treatment and, with in effect a *swarm* (the nucleus) being released into the area, along with *casts* where additional QCs are used. This *vertical* approach to mite transmission will tend to lead to a balanced relationship between mite and honeybee[84].

The approach could be supplemented by small-scale queen rearing schemes within the area to bolster colony-raising by individual beekeepers and supply other beekeepers, particularly where there is a threat of non-locally adapted bees being used.

A general comment about swarming may be useful. Beginners in particular can have a dread of finding QCs, rather than viewing it as part of normal healthy honeybee reproduction. Beekeepers should embrace swarm control, and what better way than to have nucleus boxes with 5-frames of foundation in a handy location (secured on site or in the beekeeping vehicle) and be in a position to respond at the first sign of QCs with a method such as that described above. A beekeeper who is taken by surprise and has not made any preparations will be tempted to destroy all QCs. This normally results in deferring the inevitable and is wasteful.

The approach also ensures that a number of brood frames in the parent hive are replaced each year with foundation without the need for any special replacement programme.

## Varroa-specific artificial swarm

The natural swarming or absconding of colonies in the wild has been an effective way of keeping pest levels under control. The swarmed or absconded colony gets away from an old nest infested with mites in brood cells or other organisms in the colony debris. Moreover, infected comb with brood diseases is left to the wax moth to clean up. An artificial swarm can be used to give some of this natural control.

A classic artificial swarm should be carried out in the normal way[105] (it is essentially that described in **Figure 5.2a** above, except that parent colony is moved and kept in apiary, while a full brood box is made up with the queen and put on original parent hive location with queen, the artificial swarm colony). QCs can be selected in parent colony as per **Figure 5.2a**. After seven days all unselected QCs should be removed from parent colony, one selected QC put in a nursery cage[76] and the balance distributed to mating nuclei. After three weeks, when the brood in the parent colony has emerged, two bait frames of open brood should be transferred from the artificial swarm colony to the parent colony. When sealed this brood should contain a large proportion of the mites in the parent colony and can be removed and destroyed[76]. The unmated queen in

Figure 5.2c Mating nucleus (Apidea™) Top: tools laid out, press-in cage, hive tool, pen, clear punnet and water sprayer (no smoker). Below: inspecting comb with a removed comb wedged vertically in punnet.

the nursery cage can be culled and one of the newly mated queens introduced to the parent colony.

This is an integrated Pest Management (IPM) approach to be used where mite population is high. This may happen where there is lack of solidarity within the local beekeeping area to non-treatment and non-importation of colonies. In these circumstances, conforming beekeepers can become increasingly vulnerable to the presence of DWV. In normal circumstances 'bleeding off' deformed bees in small purges over many colonies would contain the problem.

## 5.3 Managing colonies

### (a) General housekeeping

The beekeeper should set up a routine to achieve good housekeeping practice and get it established as the norm at an early stage. It does require a discipline to ensure that these *important* tasks are undertaken, as they can easily be swamped by *urgent* beekeeping work.

- keep the apiary tidy. Cut grass around stands to increase air circulation, or preferably use horticultural ground-cover fabric to act as an apron around the hive. This will also improve the visibility of bee behaviour (or dead bees) in front of the hive (**Fig. 5.3a**)
- place all brace comb, propolis scrapings and cut-out queen cells, etc., into a covered container and sort out later (**Fig. 5.3b**)
- recover wax from brace comb and old combs using a solar wax extractor (**Fig. 5.3c**)
- keep bee suit and boots clean. Gloves need to be cleaned regularly and sufficient replacement gloves should be available. Rubber gloves are easier to clean
- hive tool should be clean. Washing soda in a spray bottle is an effective way to clean the hive tool, and facilitates regular cleaning while in the apiary

The aim in keeping the apiary tidy and grass/weeds in check in the immediate vicinity is to let in sunlight and to reduce dampness. Poorly laid-out and maintained apiaries can make movement difficult for the beekeeper and lead to the loss of tools in the long grass. There is no need to expend a lot of energy cutting all the grass in a large site; it is of no benefit to the bees.

### (b) Handling colonies

Good practices should be incorporated into the approach to managing the bees in the apiary:

### Sourcing the bees

- obtain bees from a reliable source. If possible, source bees locally or raise colonies as detailed in para 5.2 above. This collective commitment within a beekeeping association or community to obtain bees within the area, can go a long way towards minimising the transmission of diseases into the area. On a wider scale, globalisation has affected beekeeping and caused the spread of many exotic parasites across national boundaries. Also, movement of bees within national boundaries due to trade in colonies or for pollination is a prime factor in the spread of diseases (para. 4.16)

- stray swarms are a potential source of disease. Pick up stray swarms, isolate them, put on foundation and feed after 3 days, subject to condition. Preferably, keep swarms close to the locations where they were picked up and check brood before introducing the colony back to a working apiary

## Specific practices

- avoid crushing bees during manipulations. It can be a common sight to see dead bees *plastered* between hive contact parts and while some of this may be unavoidable, steps should be taken to reduce its occurrence. It is good practice to use dummy boards to avoid injuring bees by rolling them when removing the first comb **(Fig. 5.3b)**. Levering out the first frame can damage the brood box allowing water to enter at the corners. In addition, working bees without gloves should be seriously considered and practiced, where possible, as it tends to bring a gentle touch to even the roughest operator! (Beginners will usually take some time before reaching the bare-handed stage) Without gloves, the beekeeper will have an awareness of brood temperature, which should help to reduce manipulating times particularly early in the season
- replacing supers/queen excluder on a populous hive can give rise to a lot of bee squashing. It requires a prudent use of the smoker as well as good technique, particularly when replacing several well filled-out supers. A common method is to put the returned super down slightly obliquely and then swivel into alignment. This can kill bees especially with a bottom beespace hive. A technique that can be effective is to put fingers below the super while standing behind the hive and gently dip the front edge onto the front of the hive. Move fingers gradually to the back and gently wriggle the super while lowering it slowly into place. Best done with bare hands or fine tight-fitting gloves. Bees are not *fools* and will move away until the super is in place!

**Figure 5.3a** Above: overgrown apiary with inadequate stands. Below: same apiary and hives, after grass and scrub have been cut and hives rearranged on raised stands that have been staggered to avoid drifting.

**Figure 5.3b** Using a dummy board avoids the squashing of bees when removing the first frame. Note, the container beside the hive to collect brace comb, propolis, etc

**Figure 5.3c** A solar wax extractor is a handy way of recovering wax for recycling from brace comb or replaced combs.

- keep hive movements to a minimum. Moving hives or migratory beekeeping can be very disruptive, and stressful to bees. Proper handling, transport and care will reduce the impact
- do not use second-hand combs or frames. Limit interchange of frames between colonies
- replace a few old combs annually in each hive to reduce contamination. Replacing all combs in the hive at the same time every few years is an alternative approach. Annual replacement of all combs is practiced by some where foul brood has been a persistent problem. Comb replacement is built into the approach to raising colonies in para. 5.2 above through the annual replacement of 3 or 4 combs.
- hive smokers should use natural (not man-made) materials as fuel, such as dry grass or well-rotted hessian sacking. Minimum smoke should be used to control the bees, and the amount of smoke should be reduced when dealing with the supers. Ideally, the smoke should be allowed to drift over the tops of the boxes in an air current
- keep colony manipulations to a minimum and have a purpose behind each manipulation. Weighing hives can reduce unnecessary colony manipulations[126,136]

## Regular inspections

- all manipulations should be carried out with a clear purpose and discipline. Routine/regular manipulations lend themselves to efficient and effective execution. Weather conditions should be suitable, and the colony should be kept under control at all times.
- observe conditions outside the hive before smoking or opening the hive. Work through brood frames at a steady pace and in an orderly fashion. The five headings in *Guide to bees and honey*[105] provide a useful structure:

1. *Has the colony sufficient room?*
2. *Is the queen present and laying the expected quantity of eggs?*
3. *a) early season. Is the colony building up in size as fast as other colonies in the apiary?*
   *b) mid season. Are there any queen cells present in the colony?*
4. *Are there any signs of disease or abnormality?*
5. *Has the colony sufficient stores to last until the next inspection?*

- work through the brood box with purpose and don't allow yourself to be distracted by "it would be nice to find out again" thoughts or always having to find the queen when there is plenty of evidence that she is there and laying. Keep the time involved to a minimum.

- close up the colony, smoke the next colony and fill up the colony record for the completed colony. The record can be structured around the five questions above (para.5.8).

## General

- be vigilant at all times and identify diseases early and always be on the look out for the abnormal. It is good practice to dedicate at least one inspection during the season to check for diseases, particularly so in the case of brood diseases. During this manipulation shake bees from each brood frame and thoroughly examine the brood
- when uniting colonies, be sure you know *why* a weak colony is weak
- be gentle when working around the colonies, walk quietly and don't knock into stands or hive parts. It can be instructive to occasionally listen to your bees in the winter period by placing your ear tightly against the brood box. Learn to pick up the low hum of a contented colony, and as a one-off to appreciate its sensitivity to disruption, drop the hive tool from a height onto firm frosty ground while listening. The colony's audible response can be a useful lesson
- take bee samples for diagnosis on a regular basis (para. 5.7). Frequent inspections should be undertaken in areas with a high incidence of disease, particularly the foul broods
- prevent robbing (para. 5.6), reduce weak colonies to 1 bee space. Seal the entrance of dead colonies
- replace supers after extraction onto the same hives for cleaning if you wish to store supers dry. Undertake this activity in late evening to avoid robbing.
- consideration should be given to refraining from or phasing out the use of chemical treatments. If used, they should be approved for disease control and should strictly follow the instructions. Be aware of current *National Regulations* and record usage of treatments (para.5.8).

## (c) Finding & marking/clipping queen

Finding and marking queens is dealt with in some detail as it is an annual exercise for most beekeepers and seems to be a source of stress and frustration for many. It can also inflict a lot of unnecessary disruption on a colony. Many beekeeping practices depend on being able to find the queen in the colony. On the other hand, as already mentioned, many beekeepers and not just beginners, feel that a routine colony manipulation is incomplete unless the queen has been seen. This is not a desirable practice as it extends the time that the colony is open, and exposed to unnecessary disturbance, increased thermal stress as well as bee deaths through crushing.

A dark queen such as *A. m. mellifera* does not stand out from the rest of the bees. Having her marked makes her much easier to find and reduces the disruption that can ensue in a situation where the queen must be found before the rest of the manipulation can continue. The subsequent presence of an unmarked queen can indicate supersedure. Also, wing clipping is often undertaken to give flexibility in swarm control (para 5.2). Beekeepers will likely try many different approaches over time and eventually adopt an approach that best suits them.

## Finding the queen

- marking of new queens raised within a colony is often undertaken as a specific task early in the following spring when bee numbers will be at a low level. However, beekeepers should have the necessary marking and clipping items on their person at all times (para. 3.4) so that the job can be done when the queen shows up in the normal course of business.

## Queen behaviour

- from the viewpoint of a beginner who has never seen a live queen, having a picture of her physical appearance is of limited use in trying to find that one insect in a box of 20,000 insects. It is most likely to be her behaviour/demeanour on the comb that allows her to be identified as she moves to the edge of the frame to get out of the daylight. A video of queen-behaviour would be a better aid in identification. In any case the beginner should make a point of getting the mentor (or an experienced beekeeper in the vicinity) to show a queen moving on a frame.
- probably the greatest deterrent to an inexperienced beekeeper finding a queen is being a bit overwhelmed, *and not expecting to find the queen in any case*. This attitude can be self-fulfilling.

## Specific queen-finding task

- in many cases the queen will be identified during the normal course of beekeeping, as mentioned above. However, for any queens not picked up in this way, a separate queen-finding task may have to be undertaken. In this case the queen will most likely be found by going through the frames one at a time. It is preferable to choose a warm, sunny afternoon when bees are flying.

■ the queen normally resides in the centre of the brood box on frames containing brood and in egg-laying mode. However, when the colony is smoked at the entrance there will be an increased likelihood of finding her on frames nearer the back of the brood box as she moves away from the smoke and the light being introduced as front frames are removed and examined. It is best to use minimum smoke during this exercise as she can be driven off the frames and onto the floor or walls, making detection much more difficult.

## Virtual enclosure

■ each frame should be held over the brood box and examined in detail. Remember that a queen will want to move out of the light. Use your eyes to initially scan round the perimeter of the frame – in effect trapping the queen in a *virtual enclosure*. Continue to move your eyes inwards in diminishing circles while being conscious all the time of any unusual movement of a bee(s) towards the edge of the frame. You can blow gently on any clumps of bees to thin them out. Turn the frame and continue checking until all the frames have been examined. In most cases the queen will turn up on one of the last frames to be examined. Be particularly alert to the possibility of a hole in the brood comb, as a queen can 'play games' with you by disappearing through the hole as you turn over the frame.

## Floor or sidewall runner

■ if the queen has still not been found it may be that you have not missed her on a frame but rather that she is a floor or sidewall runner. Young queens tend to be flighty/nervous and readily move off the frames. If this is the case it can be difficult to see an unmarked queen that is off the frames and on the floor or side walls as she will most likely be in the shade with brood frames obstructing your view.

## Second brood box

■ if you have been unable to find the queen at this stage (and there is a large and increasing population of bees), and time is running out (coming near the swarming season) a different approach is necessary. A second brood box can be set on a new floor or travelling screen alongside the parent brood box and after examination the frames from the parent can be placed in the second brood box. Placing brood frames together in pairs in the second brood box can help in finding the queen, as she will usually move to the inner faces of the brood-frame pairs to get out of the light. If the queen is not found, the walls/floor of the parent hive can now be checked. A large population of bees can make it difficult to identify a queen that may be buried in a huddle of bees in the corner of the brood box. You can blow gently on these bees to thin out the huddles. If the queen is still not detected, the frames in the second brood box can be re-examined one by one and moved back to the parent hive. If still no success, another method may have to be tried.

## Moving parent hive

- at this stage there may be a huge population of bees with several supers. Choose a good day when the bees are flying freely. Move the parent hive to a new position in the apiary as far away as possible from the original stand position. Leave a super (or the supers that were on the hive) behind on the original location. Wait for at least 30 minutes until most of the flying bees have returned to the original stand and then smoke the colony gently. Now go through the frames in the parent hive as you would for a normal queen search as above. If queen is not yet found, resort to using a second brood box. Most queens should be found in this way.

## Filtering

- an approach using a queen filtering box (a brood box with a queen-excluder on the floor) by shaking all the bees off the frames in the colony brood box and then smoking the bees down through the bottom of the filtering box. The queen should be found on the bottom of the filtering box. This is disruptive to bees and really only justified where there is a very aggressive colony that requires the queen to be found and replaced.

## Overview

- if you generally have difficulty finding queens, don't delay in getting help instead of putting your bees through a lot of disruption. Aggressive and runny bees can make things difficult.  Also, from personal experience, a colony can swarm out early, before the first queen cell is sealed, if subjected to excessive handling[105]. In the end you can resort to using a method of carrying out an artificial swarm without finding the queen[211]. This would be preferred to using the filtering approach above unless of course there is an unsuitable queen (particularly aggressive) that must be removed.

## Catching and marking/clipping the queen

■ The queen can be marked using colours from the international colour code below or a single bright colour such as white. Years ending in 1 or 6, white; 2 or 7, yellow; 3 or 8, red; 4 or 9 green; 5 or 0, blue. It is necessary to 'restrain' the queen in some manner in order to mark or clip. This will involve either catching the queen by hand or trapping her in some other way.

## Using a 'plunger-type' marking cage

■ Allow the queen to pass into the tube, with workers if necessary, making sure that she does not get injured against the hard edge of the tube (**Fig 5.3d**). Insert the plunger and move gently upwards allowing the workers to escape through the slits at the top. It is important that a properly fitted sponge is in place to ensure the correct pressure on the constrained queen. Align the Q along one slit and press gently to keep her in position (sufficient pressure only to keep her from wriggling). Use the felt-tipped colouring-pencil (that should be kept in the bee suit pocket) to put a small daub of paint in the middle of the thorax. Allow paint to dry before releasing the queen. If you are also clipping the queen, gently release one wing through the slit and clip off up to half of the wing top. Release the queen onto a comb with brood and check that her conformation is good, and that she has settled back normally into the colony. Smoke colony gently, return/adjust frames and close up.

## By hand

■ Working on your own, constrain the queen temporarily on the comb with the press-in cage **(Fig. 5.3d)**. Use water sprayer to remove sweat from the hands to make them more neutral for the odour-sensitive bees. You may use gloves up to this point, to avoid contact with propolis, and can remove them now and water spray to remove any sweat. Ensure marker pen and scissors are at hand. Remove the press-in cage and catch the queen by the thorax with index finger and thumb of the right hand and transfer to the left-hand finger and thumb with the middle finger supporting the abdomen **(Fig. 5.3e)**. Put a daub of paint on the centre of the thorax and while the paint is drying clip the wing, making sure that the queen's legs are well clear. Release the queen back to the colony as above.

 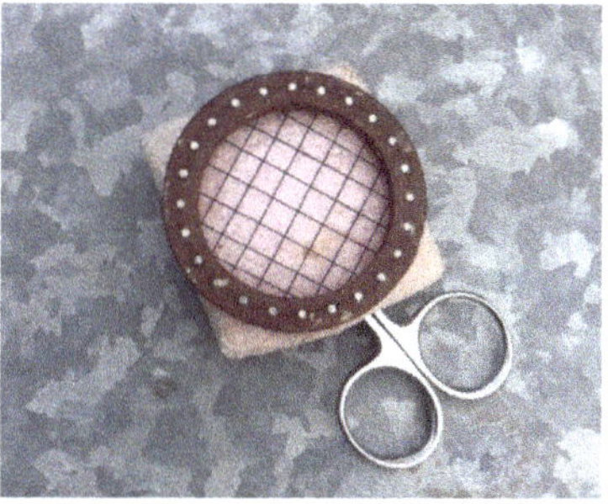

Figure 5.3d (L): plunger marking cage; sponge at the top end of the plunger should be clean, and free from honey and propolis. It can help to regulate the sponge pressure on the queen if a sponge/elastic band is fitted to the base of the plunger. (R): press-in cage ('crown of thorns') in block of polystyrene with pair of blunted scissors also pushed in. Note: these items should be kept in your bee suit pockets and removed when required without taking your eyes off the brood comb.

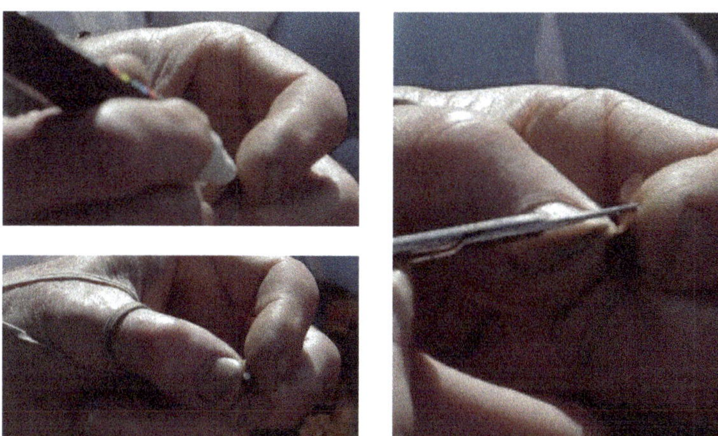

Figure 5.3e Left, top and below: queen being held by thorax and marked with white pen. Right: wing being clipped. NB Ignore author's dirty nails!

## 5.4 Disinfecting and sterilizing

Beekeepers should employ a strict regime for disinfecting and sterilizing equipment and tools.

### Hive parts

Wooden hive parts such as brood boxes, crown boards and floors should be scraped to remove brace comb and propolis before being reused. The groove along the frame runners requires particular attention. The hive parts should then be subjected to a thorough scorching using a blowtorch to boil off the remnants of propolis and to singe the wood to a coffee brown shade **(Fig. 5.4a)**. Plastic runners can be shielded from the flame using a wide painter's scraper. Polystyrene parts should be scraped with care and washed in a hot solution of washing soda.

### Frames

All frames to be re-used after old comb has been cut out should be disinfected. Washing soda is cheap and widely available and can be used in a solution of 100g to 1 litre of water (1lb to a gallon). The wooden frames should be boiled in the washing soda solution for 10 minutes, brushed clean of wax and propolis and rinsed in water **(Fig 5.4b)**. Washing soda is corrosive and the instructions on the product should be followed.

**Note:**

- all wooden frames in a hive that has any comb infected with AFB must be burned (para. 4.1)
- to facilitate the small-scale beekeeper, beekeeping groups or associations should consider having a suitable water boiler for hire, as is common in the case of honey extractors
- using new frames rather than rejuvenating old ones may involve little extra expense
- burning all the contents of the affected hives on site after an outbreak of AFB or EFB and scorching the wooden parts is common practice. As an added precaution, some beekeepers also use sodium hypochlorite (bleach) at a concentration of 0.5% and an immersion period of 20 minutes as pioneered in New Zealand[94]. Bleach can also be used to sterilize polystyrene hives, as they cannot be blowtorched.

## Brood and super boxes with comb.

Before being reused, sterilize brood comb with acetic acid, but acetic acid should not be used for AFB-infected and EFB-infected comb. (Replacing old brood comb with foundation, after disinfecting the frames, should be considered rather than storing a brood box with old, sterilized brood comb over winter.) Combs in supers do not normally need to be sterilized before reuse. An 80% strength of acetic acid should be used, and if the solution is being made up from glacial (100%) acetic acid, the acid should be added to the water and not *vice versa*. Remember that it is corrosive, and strict precautions should be followed to avoid the inhalation of fumes or getting the acid on skin by using masks and gloves:

- the brace comb and propolis should be first scraped off. Bare metal parts such as frame runners should be coated with petroleum jelly
- the boxes with frames should be put outside on a solid board or hive floor and a saucer with an absorbent pad put on top of the frames
- 120 ml of 80% acetic acid should be poured onto the pad, an eke added on top and covered with a crown board that has the feedholes sealed
- joints should be taped and a roof put on top. Alternatively, this arrangement can be made up inside a large black plastic bag (e.g. "wheelie" bin liner) and the liner knotted on top. This will have the advantage of providing a good seal and being black will absorb heat in daylight and raise temperature in cool weather conditions (**Fig. 5.4c**)
- after seven days in normal summer conditions, the boxes should be set aside and aired for a few days before being used.

## General cleaning of tools and other items.

The beekeeper's toolbox and its contents should be disinfected on a regular basis. A solution of 100g of washing soda to 1 litre of hot water (1lb to a gallon) will remove propolis from items such as hive tools, scrapers, smokers and rubber gloves. In some cases, the items may need to be soaked in the washing soda solution. Hive tools and scrapers can be blowtorched. Beekeeping garments should be cleaned regularly with some washing soda used in the wash to remove propolis.

**Figure 5.4a.** Scorching brood box. Use strong flame, work away from you, and ensure that wood is charred. Scorch into crevices and protect plastic runners from flame using a broad scraper as a heat shield.

**Figure 5.4b** Brood frames being brushed after immersion in a boiling solution of washing soda

**Figure 5.4c** Sterilize brood combs with acetic acid before reuse (not for AFB/EFB). Use eke to give headroom, replace crown board and roof. A good seal will be obtained if hive is assembled inside a black "wheelie" bin liner and the roof replaced (inset).

## **5.5** Nutrition and feeding

Adequate nutrition is critical for the health of honeybees. Honey (and nectar) along with pollen provides carbohydrates for energy requirements as well as protein, minerals, vitamins and fats for their glandular and body development. Water assists in stores dilution and colony temperature control **(Fig. 5.5a)**.

The beekeeper should ensure, as far as possible, that colonies are located in areas where adequate levels of nectar and pollen forage are available[150]. In the wrong location, it is an uphill battle to keep healthy bees.

Nutrition is necessary at three levels: the colony, adult bee and larval levels[32]. Larvae and young adult bees are particularly dependent on adequate supplies of food, pollen being necessary for their proper glandular development. A good supply of pollen is important for the winter bees to see them through to the spring. At the colony level, a large supply of stored honey is necessary to meet their energy needs over the winter period when no fresh nectar is available. It has been estimated that in temperate regions in the US, for thermoregulation, a small colony will use on average 0.42 kg (~1 lb) per week. However, stores consumption will rise to an average of 0.84 kg (2 lb) per week when brood rearing is present[186]. From experience, north European dark bee (*Apis mellifera mellifera*) colonies will winter with not more than half of the US stores use **(Fig 5.5b)**.

Feeding bees to replace honey stores is now an established part of modern beekeeping[139]. Getting it right will help to ensure that strong colonies are maintained and that the risk of robbing is kept to a minimum. In addition, weak colonies make the bees vulnerable to attack by pathogens and parasites. Generally, there should be no need to feed pollen or pollen substitutes if the apiary location is chosen with care. In regions where large-scale monoculture crops, severely trimmed hedges and weed removal from headlands are common, the beginner-beekeeper in particular should be wary, as the bees may have to be nursed along in a bee-hostile environment. In these situations, some beekeepers provide a lot of the bees' food supply, including protein supplements, in their home apiary rather than moving the bees to a location with adequate forage. This trend has the appearance of the earlier movement from "free range" to "deep litter" or "battery" hens.

### Reason for feeding:

Beekeepers should not lose sight of the fact that honeybees strictly control temperature and convert chemical energy (stores) to heat energy to maintain high brood nest temperature even during late winter conditions. In regions with cool damp climates, the health of the honeybee colony depends on access to stored food. As well as the need to ensure adequate winter stores there are many other situations where colonies should be fed[108].

In general, feeding should be undertaken at any time of the year when bees are low in stores, and this would normally mean less than 5 kg (10 lb). In poor foraging weather, colonies can be lost through starvation, and this is particularly so in the case of the parent hive after an artificial swarm manipulation unless immediate feeding is undertaken. Feeding can be used to help new colonies to draw out wax comb and build up the bee population. Stimulative feeding to encourage a rapid build-up for specific purposes can be undertaken but this practice is declining.

## When to feed:

Winter stores should be given at the earliest opportunity. Some prefer to wait for late autumn flows, such as the ivy flow and then top up later with sugar syrup. Of course, the flow may not yield or be insufficient. The beekeeper is then left with getting the bees to ripen sugar syrup in the month of October or later at a time when cold, damp conditions are common. This is a perfect recipe to stress the winter bees that are an essential part of the spring build up. Feeding until the required weight of stores is in the hive by the end of September would seem to be prudent. The minimum level of stores can be in the comb and sealed by this time, and if the ivy flow comes later it is a bonus. With adequate stores in the hive, there should be no need to feed in the spring.

## Quantity of feeding:

Some beekeepers feed their bees early until they can take no more and end up after a late-autumn flow with a stores-bound colony in the spring. How much stores does a colony really need? This will depend on many factors such as the strength of the colony, climate, time of first honey flow and bee strain. Each beekeeper needs to establish what is acceptable in their circumstances. A typical winter stores use (depicted by "w") for a medium-strength colony in an area with a spring flow **(Fig. 5.5b)** is 20 lb (9 kg) from the beginning of October to the start of the spring flow in the following April. Allowing for a 10 lb (5 kg) buffer, a weight of 30 lb (14 kg) at the beginning of October should be sufficient for a medium-strength colony. Corresponding estimates are 40 lb (18 kg) for a strong colony and 20 lb (9 kg) for a weak colony and nucleus. The hive in question is the

"Modified Commercial" and the bee strain is the native North European dark bee, *Apis mellifera mellifera,* which has a reputation for being frugal in terms of winter stores usage.

How much do we feed? Prior to feeding, the amount of stored honey can be estimated from the equivalent number of frames of sealed honey. This is time consuming and involves disturbing the colony when robbing is prevalent. The usual practice of hefting from the back of the hive is crude and the effort required is dependent on many factors, not least the type of hive stand, as even a heavy hive resting on a narrow stand will always feel light when hefted from behind. With experience the accuracy can be improved but there will still be a big bandwidth of uncertainty. There is only one way of doing this work with a reasonable degree of accuracy and that is by weighing the hives and making allowance for the hive and accessories[108,146,2]. A simple lever method can be used[126,136]. The method is cheap, accurate, stable and simple to use **(Fig. 5.5c)**. As well as ensuring the accurate provisioning of winter stores in the hands of the small-scale beekeeper in particular, weighing can provide interesting possibilities: facilitating ongoing monitoring of winter stores (and honey crop), supering, identification of robbing and permitting reduced manipulation.

## Feeding options:

Stored frames of sealed honeycomb represent an ideal feed particularly for weak colonies or nuclei, but these are usually in short supply. The frames of stores should only come from one's own apiary and be disease free. There should be no need to disturb colonies to feed them in mid-winter if proper feeding has been carried out in the autumn. If necessary (and in the absence of weighing there will be a greater level of uncertainty), colonies can be fed fondant.

Sugar syrup made from white granulated sugar is the usual feed and many different mixture strengths and methods of preparations are in use. With autumn feeding, a strong mixture of white sugar in water is necessary as this feed will be stored. It is here that the term 2:1 crops up in much of the beekeeping literature, but this has meant different things to different people over the years. It is sometimes used as a true ratio of two parts sugar to one part water by weight, and this gives a saturated solution of sucrose (66.7%) at 20°C which can result in granulation and clogging the mesh in the lid of a contact feeder (solubility of sucrose in water at 20°C is 2.04 kg per 1 litre of water). The 2:1 ratio is more usually referred to as 2 lb of sugar in 1 pint of water giving a 61.7% sugar concentration. This term is confusing particularly for beginners to beekeeping as it is not a normal ratio of weight-to-weight or volume-to-volume and relates to units that are officially about 50 years out of date in Europe. Using current units, 4 kg of sugar added to 2.5 litres of water gives 61.7% concentration or alternatively, adding 4kgs to a 5 litre bucket and making up to 5 litres with water also gives the same concentration. It can be seen from the foregoing that the same sugar concentration will be obtained by filling any container exactly half full with water and adding sugar to make up the full container of mixture. For spring feeding (stimulative feeding) or feeding bees for immediate use the old term was a 1:1 mixture usually meaning 1 lb of sugar to 1 pint of water giving a 44.4% concentration. A better 1:1 ratio is 1 kg of sugar to I litre (1 kg) of water giving a 50% concentration which can be metabolised directly by the bees. Alternatively add 3 kg of sugar to a 5-litre bucket and make up to 5 litres with water giving a 49.0% concentration. All of the above mixtures can be made up using boiling water and stirring until dissolved, but do not boil the mixture.

To summarise:

- a **"4-strength"** mixture, a strong feed (62%) for storage: put 4 x 1 kg bags of sugar into a 5-litre bucket and make up to 5 litres of syrup[139].
- a **"3-strength"** mixture, a weak feed (49%) for immediate use: put 3 x 1 kg bags of sugar into a 5-litre bucket and make up to 5 litres of sprup[139].

To avoid fungal growth in the syrup that may appear after a period of time, a few drops of thymol can be added to the mixture. A small bottle, one-third filled with thymol crystals and topped-up with ethanol (or surgical spirit) will go a long way by using 2.5 ml (half a teaspoonful) to 5 litres of syrup.

## Delivery system and precautions:

Winter feeding is normally carried out using rapid feeders (Miller type) or bucket contact feeders **(Fig. 5.5d).** For the small-scale beekeeper, the contact feeder can be used both to mix the feed and to give it to the bees, and as long as the mesh is not blocked the feed will be taken down quickly. Frame feeders can provide a simple method of feeding small colonies or nuclei and removes the need for over-brood box space. They also appear to reduce the level of robbing **(Fig. 5.5e).** Floats need to be provided, and the feeder can be removed when feeding is complete.

Mid to late winter feeding of hives or nuclei, if necessary, can be undertaken using fondant placed over the feed openings in the crown board. Using large freezer bags with a zip, over 2 kg (4 lb) of fondant can be held. Specially prepared "fondant" products in plastic bags are also commercially available.

Colonies should be fed late in the evening with all colonies being fed at the same time, entrances reduced and spilling of syrup avoided. Remember it is much easier to prevent robbing starting than to stop it later.

**Figure 5.5a** A brood frame containing honey and sugar-syrup stores (dark cappings) on top, and a variety of honey-glazed pollen going into winter.

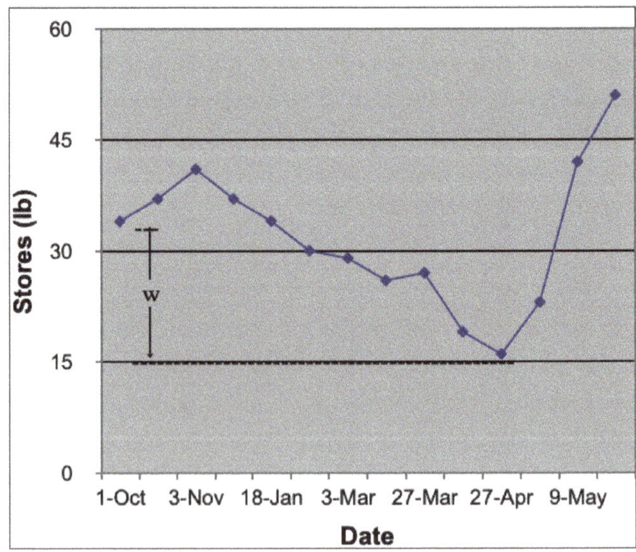

**Figure 5.5b** Typical stores usage pattern for a medium-strength colony over the October to May period.

**Figure 5.5c** Weighing a hive on a stand that has been adapted to ensure that the inserted lever (short plank of wood) is parallel with the floor of the hive when the side of the hive is just lifting off the stand.

**Figure 5.5d** A contact feeder over the feed hole in the crown board. A super is used to provide headroom.

**Figure 5.5e** A frame feeder can be a simple way to deliver sugar syrup particularly to a nucleus where headroom may be restricted, and it also appears to reduce robbing.

## 5.6 Managing robbing

Robbing presents dangers for both the robbed and the robbing colonies. The robbed colony may be driven to collapse or at the least may become stressed, which can have a detrimental effect on its health, particularly if it coincides with other conditions and parasites. The robbing colony may contract disease, mainly through infected honey, and colonies may exchange diseases by contact.

Honeybee colonies in these islands are generally only robbed by other honeybees and wasps. Ants, woodlice and earwigs can be found in hives usually, in the roof space, but seldom present a threat. Occasionally ants can target a small colony in a mini-nucleus, such as an Apidea™, but usually it is not a problem.

Robbing is a major vector in spreading diseases[81] and must be controlled to ensure the long-term health of the bees. It is probably the single most important factor in the spread of AFB[94]. See paragraph 4.16, disease transmission.

It can be difficult to detect robbing. There may be fighting at the entrance or bees may be flying erratically. Sometimes there may be *silent* robbing, with the colony appearing to behave normally, but an experienced beekeeper should be able to detect from the bees' landing behaviour that some bees are coming in light and going out loaded. We have all been fooled as beginners when feeling gratified at the sight of a lot of activity in a colony in early spring and thinking that all was well, we later discovered that the colony was being robbed out.  If you are not present in the apiary when the bees are flying, it can be difficult to detect robbing. An unusual level of cappings below the brood box and an untypical reduction in weight (if you weigh your hives) can point towards robbing.

Since both the robbed and robbing colonies are under threat from a health viewpoint, this is a situation where members of the beekeeping community are dependent on each other. There is an individual, as well as a community-level, interest in robbing. At the individual level, having strong colonies by itself will not protect the beekeeper, as such colonies are more likely to be the robbers and be exposed to weak or diseased colonies in their neighbourhood. At the community level, the aim should be to reduce the level of inter-apiary robbing and also to reduce the impact of robbing on colonies, both robbed and robbing.

In all cases the intent should be to reduce the likelihood of robbing starting in the first place. Remember it is easier to prevent than stop robbing after it has taken hold.

At the individual beekeeper level:

- keep strong colonies and build up nuclei
- act swiftly on queenless colonies, particularly early or late in the season
- adjust the entrance to size depending on the size of the colony and time of year and do so in good time
- reduce spillage of sugar syrup or honey and feed in the evening (para. 5.5)
- keep manipulation time to a minimum when there is no nectar flow, and preferably carry out manipulations only in the evenings after mid-summer
- ensure that hives are bee tight
- take precautions when using varroa treatments that have a strong odour, as these can mask colony odour and confuse guard bees.  Small colonies are particularly vulnerable to this effect.
- close up dead colonies immediately
- detect disease early and provide urgent remedial action.


At the community level:

- encourage all beekeepers in the locality to join a beekeeping association and promote a co-operative spirit
- improve standards of beekeeping in the area
- discourage high local densities of apiaries
- distribute information on local diseases and praise early detection and disclosure, particularly of notifiable diseases. There is no shame in having a disease, but there is shame in having it and not knowing it or knowing and doing nothing about it.


If robbing starts it can be very difficult to stop. Where it has become well established, it can be difficult to avoid sacrificing a colony. In some cases, it may be necessary to move the robbed colony. All entrances should be reduced or stuffed with grass. Little has changed over the years with regard to the problem of robbing; in the early 18th century robbing was one of the major problems encountered and Irish beekeepers were advised, "… to lessen the passages of all your hives in the beginning of August to about half an inch in breadth …    ."[196]

The common wasp (*Vespula vulgaris*) can devastate colonies and is widespread throughout Britain and Ireland while the German wasp (*Vespula germanica*) is also present. Where large colonies of wasps are in the vicinity of an apiary, they can threaten bee colonies from late summer. Small or queenless colonies are particularly at risk and can be overwhelmed, with all honey and even brood removed within a short period. For this reason, it is necessary to act in good time, be alert to the level of wasp activity in the neighbourhood and take the

necessary precautions **(Fig. 5.6a)**

**Figure 5.6a** A colony with grass stuffed into the hive entrance, under threat from wasps in the late summer.

## **5.7** Detecting disease/sampling/post-mortems

With so many maladies of honeybees present in current beekeeping, knowing the health status of our colonies is an important part of honeybee management. During normal colony manipulations beekeepers should be vigilant and identify abnormal conditions early. In this regard beekeepers should be familiar with the signs of diseases in Part 4. In the case of American and European foul broods (para. 4.1, 4.2), specific brood checks should be undertaken. Furthermore, taking representative samples from hives for testing should be an ongoing part of present-day beekeeping. Before taking samples check for any special requirements of *National Disease Policy* (para. 4.17)

### Detecting the presence of disease

The beekeeper is normally interested in getting some measure of *prevalence*, namely the proportion of the total colony population that has the disease.  A fuller understanding of the virulence of the disease can be obtained from a knowledge of the parasite *load* or *intensity,* which is the average number of parasites per bee present in the bees that have the disease. To detect the presence of disease, the adult bees should be sampled at an age when the disease would be expected to have expressed itself. The resulting infestation levels would generally be higher than the population average.

The number of bees in a sample is generally a trade-off between accuracy and the time taken to collect and test the sample. Normally a sample of about 30 or more bees properly collected gives sufficient accuracy for beekeeping purposes.

### Where to take the bee sample

As a general rule, samples of flying bees should be collected, as these will normally be over three weeks old, and this usually means collecting bees from the outside of the hive. In some cases, beekeepers have found it more convenient to collect from inside the hive using, for example, a glass jar over a feeding hole in the crown board. However, parasite prevalence and load will be lower for conditions such as nosema and tracheal mites, and recent research has shown that sampling inside the hive for a condition such as CBPV can fail to give positive results for an infected colony[30].

## How to take the bee sample

Collecting flying bees from the landing board is the preferred method, and if necessary, the entrance can be closed for a few minutes to allow sufficient bees to accumulate. A sample taken after a colony manipulation is best avoided as young bees can drift outside the hive and be included. The sample can be collected using a matchbox that is about two-thirds open with the bees being *combed* from the landing board into the box. It is difficult to gauge by weight when the matchbox is full (~ 30 bees weigh about 3 gm). This is not helped if gloves are worn, which may be necessary if bees are in defensive mode late in the season. If a matchbox with a window is used **(Fig. 5.7a)**, sufficient bees will be collected when the bees do not move freely in the box when closed. Several samples in matchboxes can be taken using one box cover with a window by pushing the box of bees into a new cover. Matchboxes should be pre-labelled with colony numbers. (The matchbox with a window is also useful for the temporary holding of a queen). If necessary, a sample can be taken from the outside frames in a hive. Avoid drones, ensure that the queen is not on the frame, and *comb* the bees off the frames into a matchbox as before. The bees can be killed by placing matchboxes in a domestic freezer for 24 hours.

In most cases, the samples will be of live bees. There are occasions when only dead specimens will be available. This will happen in the case of poisoning, and a sample size of 200 bees should be collected. If bees have been dead too long, the samples may not be suitable.

## How to take a comb sample

A piece of brood comb, 15 mm x 15 mm, should be used. It is essential that cells with nectar, stored honey or pollen be avoided. The brood sample should have a mix of open and sealed cells, where available, and should include cells that are suspicious. Wrap sample in a sheet of newspaper, place in a plastic bag and seal with duct tape/staples. Colony number should be attached to sample.

## Labelling samples

All samples should be clearly marked and accompanied by a fully completed submission form (where required); name, address, contact details (phone/email), colony number, location etc as specified by the local bee authority.

Samples should be sent with minimum delay to the local bee diagnostic centre (refer to *National Disease Policy*, para. 4.17**).**

**Figure 5.7a** Taking a sample of bees by combing the bees off the landing board into a matchbox with a window (inset).

## Post-Mortems

In the first instance all 'die outs' will tend to look much the same, since no matter what the cause the colony will normally end up with a reducing population, temperature stress, dead brood/bees and eventually mould. An illustration of how this can happen is demonstrated for a tracheal mite (acarine) infested colony, and the *Mortality Spiral* is shown in **Figure 5.7b**. It is important to be familiar with the disease conditions in Part 4 and to observe, take notes, photographs, samples and keep an open mind. Finding a 'die out' should not be the first indication that there is a problem. There will normally be early warning signs during manipulations and routine apiary visits.

**Outside the hive:** Large numbers of dead and dying bees piled together would indicate poisoning (see para. 4.13f) while in the case of disease there is likely to be deaths over a period of days with bees in various stages of decay. CBPV (para. 4.9) and DWV (para. 4.11) could be indicated with bee condition and any live bee behaviour, providing important information. Tracheal mite infestation (para. 4.5) may have dead and dying bees on hive, stand, ground or on stalks of grass. White or grey brood dummies would normally point towards chalk brood (para. 4.3). The debris pattern on a screen insert over time should also be an indicator of colony condition, particularly when compared with others. Reducing amounts of debris and fragmented  clusters of debris would flag a problem. However, an abnormal amount of cappings could indicate robbing. Dysentery would cause soiling on the landing board or near the entrance and could indicate nosema or tracheal mites. Hence, observing the situation outside the hive even before opening may tell a lot about colony health[196a].

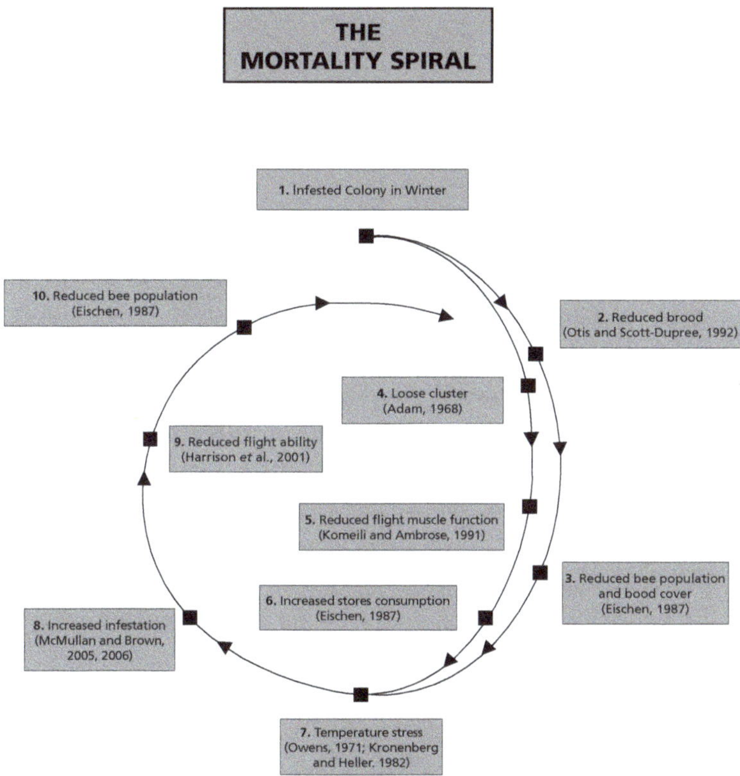


**Figure 5.7b** A colony in terminal decline, showing the dynamic influences on a tracheal mite (acarine) infested colony during the winter/early spring period. (McMullan and Brown, 2009)[134]

**Inside the hive:** Some of the bees found lying outside the hive may also be found inside. Brood comb needs to be examined for all disease conditions (Part 4). Patches of infected larvae, as opposed to infected individually isolated cells can indicate that infected (robbed) honey has been the cause. Queen conditions and the presence of isolation starvation **(Fig. 4.13a)** may also be indicated. In summer conditions the state of the comb may deteriorate quickly. However, in the case of long-dead colonies infested with tracheal mites, in cold winter conditions even where all the flesh is gone, mites at all stages can be left intact, "preserved" in the tracheae[138].

## **5.8** Records

### (a) Colony record

Colony records should fulfil a number of needs. They should give an account, under a number of headings, of the observed state or health of the colony at a point in time. Over time, this will allow the beekeeper to track the development of the colony and be a critical aid in its management.

The records kept by a small-scale beekeeper are likely to be different from those of commercial beekeepers. The small-scale beekeeper can usually afford to keep more detailed records reflecting their closer relationship with individual colonies. When starting off we have all kept detailed notes, sometimes in A4 pads, with no real structure involved and detail that was seldom looked at again. Using a tabular record, however, provides a checklist and requires minimum writing. The columns can be ticked or numbers added, and this can be undertaken well within the "two-minute" interval after smoking and before manipulating the next colony **(Fig. 5.8a)**.

At the end of the day, record keeping is personal to the beekeeper, and the headings used should have the information that is needed to manage the colonies. The approach to manipulating the colony, using the five questions[105] described in paragraph 5.3b, permits a useful structure against which to report. However, with the advent of *Varroa* mites and the use of "animal" treatments, certain key records including batch numbers of remedies etc, will be required within the European Union where colonies are still being treated. An example of a colony record that requires minimum writing is given in **Figure 5.8b** and makes provision for recording mite drops and hive weights.

### (b) Medicines record ("Animal Remedies Record")

All countries within the European Union are required to conform to certain regulations to control the use of animal treatments or remedies. The requirements are laid down in European Communities (Animal Remedies) (No. 2) Regulation 2007, (S.I. No. 786 of 2007) as amended by European Communities (Animal Remedies) (Amended) Regulations 2009, (S.I. No. 182 of 2009). This legislation stipulates that records be kept on the purchase and administration of medicines.

Under new EU animal health law that is expected to come into force in 2021, all apiaries will be required to be registered for honeybee health reasons, in addition to the present requirement to register as a primary food producer. All legislation is subject to amendment and beekeepers need to keep up with the current *National* requirements in their country (para. 4.17). In Ireland, the legislation is under the Department of Agriculture, Food and Marine (DAFM); https://www.agriculture.gov.ie

Record keeping requires that:

The owner or person in charge of a food-producing animal must keep at his or her premises a record ("Animal Remedies Record") of all animal remedies purchased and administered. This record must be kept for ALL food-producing animals including honeybees, for five years.

Records that must be kept for all animal remedies purchased and administered (as set down in Regulation 42 of SI No. 786/2007):

*Purchase/incoming details:*
- *Quantity*
- *Authorised name of the animal remedy*
- *Date of Receipt*
- *Name and address of supplier*

*Administration/Outgoing details:*
- *Date of Administration*
- *Authorised name and quantity of the animal remedy administered*
- *Identity of animal to which the animal remedy was administered*
- *Date of expiry of a withdrawal period*
- *Name of person who administered the animal remedy*
- *Name of prescribing veterinary practitioner (if applicable)*
- *Quantities of unused or expired animal remedies which were returned*

The beekeeper can meet the EU requirements by holding the invoices for the treatments purchased (Purchases/incoming details) and by keeping a Record of Administration (Administration/outgoings), such as the one in **Figure 5.8c.**

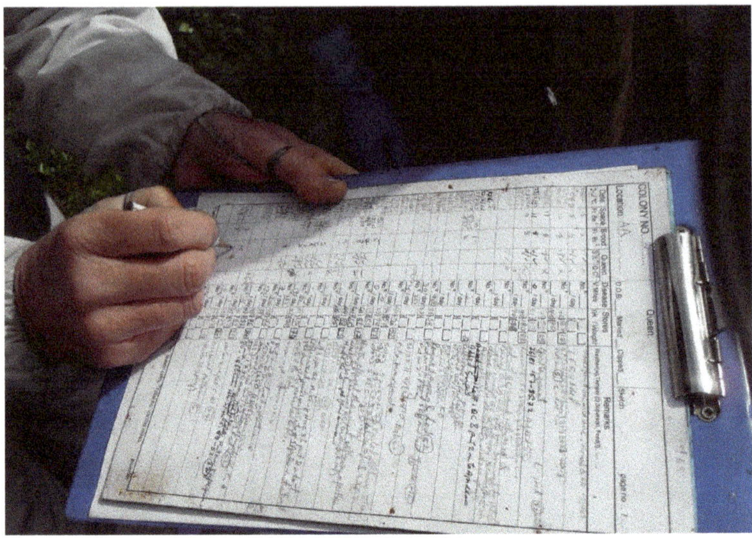

**Figure 5.8a** Filling up record after colony manipulation, and during the two-minute wait after smoking and before manipulating the next colony.

| COLONY NO. ......... Location: ......................... | | | | Queen: D.O.B:__Marked:__Clipped:__ | | Page no.... /.... Sketch: |
|---|---|---|---|---|---|---|
| Date | Space | Brood | Queen | Disease | Stores | Remarks |
| -- / °C | Fr. Be.[1] | Fr. Br.[2] | S[3]/L[4]  Q.C[5] | V.Mites | ok / Weight | weather(w),temper.(t), supers(s), feed(f),.......... |
| | | | | No.[6] __  __/ day | } __ [_] }{  ] | |
| | | | | No.[6] __  __/ day | } __ [_] }{  ] | |
| | | | | No.[6] __  __/ day | } __ [_] }{  ] | |
| | | | | No.[6] __  __/ day | } __ [_] }{  ] | |
| | | | | No.[6] __  __/ day | } __ [_] }{  ] | |

[1]Frames of bees; [2]Frames of brood; [3]Queen seen; [4]Queen laying; [5]No. queen cells; [6]Total no. varroa mites.

**Figure 5.8b** A sample "Colony Record" that has designated columns, with a row to record each manipulation.

**RECORD OF ADMINISTRATION (Animal Remedies)**

Name of  administrator (beekeeper): ___________________
Name of veterinary surgeon (if applicable): ___________________

| Date of Admin. | Name of Product | Quantity Admin. (Qty Unused) | Withdrawal Period (Days) | Batch No. | Identity of Hives |
|---|---|---|---|---|---|
| | | | | | |
| | | | | | |
| | | | | | |
| | | | | | |

Figure 5.8c A sample "Record of Administration" to comply with the European Communities (Animal Remedies) regulations.

# PART 6 | Forage and Honey

## 6.1 Bee forage

The health and abundance of honeybees depend on the availability of nectar and pollen[218] (as well as propolis and water). In Ireland this is largely provided by the natural habitat. Although habitat loss is an increasing problem, Ireland still retains a wide variety of flora and with actions being undertaken through the Pollinator Plan (para. 2.4), will hopefully be able to stop the decline and restore the availability of forage. A good all-year round supply of nectar and pollen-bearing plants is essential for honeybees as well as other pollinators. While this is normally available in urban locations it cannot be assumed for other locations. In country areas there is a need to ensure that apiaries are properly located where there is a good spread of forage types within reach. Beginner-beekeepers can sometimes be deceived by the presence of lots of trees in their neighbourhood, only to discover that some are of little use to bees, such as ash trees, unless they are supporting ivy. Sectoral guidelines developed for the implementation of the Pollinator Plan[148] include one specifically designed for farmers, "Farmland: actions to help pollinators". Actions such as maintaining natural *flowering* hedgerows, allowing wildflowers to grow around the farm, providing nest places for wild bees, minimising artificial fertiliser use and reducing pesticide inputs are recommended.

In country (rural) areas the early-flowering pollen plants such as gorse (furze/whins), and blackthorn are of particular importance (**Fig. 6.1a**). The typical colour of the pollen is inset in each of the forage figures[111]. Others include hazel and willow. The first important nectar flows usually come from dandelions **(Fig. 6.1a)** and this normally represents the first weight gain by the colony where it is not unusual to have hive weight increases of up to 1kg (3lb) per day in North County Dublin during a warm spell in late April. In May, sycamore trees usually offer a reliable forage and bees can often be seen still foraging on the blossoms during light rain showers **(Fig. 6.1a)**. Towards the second half of May the hawthorn can give good flows, but it is a fickle plant and bees will seldom be seen working on it **(Fig. 6.1b).** White clover, the old reliable in former years, can still flow well particularly in warm humid conditions from mid-June to the end of July. Even in recent years it has been known for yields up to 4 kg (10 lb) per day in old–lea clover in County Galway **(Fig. 6.1b).** Bramble (blackberry) is usually reliable and in recent times appears to have a longer flowering season (**Fig. 6.1b**). Going to the heather can extend the beekeeping year but it can also be a fickle crop. Bell heather (**Fig. 6.1b**) is usually more reliable than ling heather (**Fig. 6.1c**) but a lot depends on location. Ling on the Dublin Mountains appears to be more reliable than on some of the raised bogs in the West. Beekeepers should be mindful that bogland is a special habitat that has taken thousands of years to evolve and should treat it with care especially when using vehicular traffic. Its flora facilitates the bogs' role as a carbon sink, contributes to biodiversity and also enhances the natural beauty of the countryside **(Fig. 6.1d)**. Being in flower later in the year than bell heather, cool evenings can create extra work for the bees in ripening the nectar from ling. Insulation-wrapping the hives can help **(Fig. 6.1e)**.

However, when it comes to reliability, the ivy flows in recent years have been prolific in many parts of the country **(Fig. 6.1c)**. In North County Dublin increases in stores of up to 15 kg (35 lb) between September and November are commonplace while the plentiful supplies of nectar and pollen contribute to large increases in those valuable 'winter-bees'. Watching the bees bringing in pollen from the ivy to build up fat bodies to sustain them through to the springtime is always a delight for beekeepers (**Fig. 6.1f**). The same satisfaction is present when pollen-watching in early spring, and seeing a colony survive the winter and building up its fresh pollen stores. The downside to these strong ivy flows is that our frugal dark bees will not consume all these extra stores and colonies can become 'stores bound' in the spring. However, a few frames of ivy honey (slabs of lard!) when removed and stored can become very useful later on to make up nucleus boxes.

Of course, not all bee forage is from the wild environment. Honeybees are invaluable in pollinating some food crops, and some beekeepers engage in migratory beekeeping and move their colonies to these areas. Oilseed rape **(Fig. 6.1c)** and soft fruit (raspberries and blueberries) as well as top fruit such as apples **(Fig. 6.1c)** benefit greatly from honeybee pollination resulting in better yields and higher quality. Recent estimates put a high value on bramley apple pollination in Northern Ireland[207] as well as for oilseed rape in the Republic of Ireland[195].

As already mentioned, forage in urban areas is more reliable with blossom being available for most of the year. This is reflected in a recent study in the UK that shows honeybees in more rural locations had lower colony strength, less rich stored pollen and higher incidence of *Nosema* sp. throughout the season[178]. While most of the 'country' plants above are present in urban areas, a higher proportion of other plants are visited by bees. The urban plants in this context are 'garden plants' and of course will also be available in individual gardens dotted throughout the countryside. The following garden plants are important: ericas, viburnum, crocus, dandelion, horse chestnut, cotoneaster, lime and hebes (**Fig. 6.1g**). Sectoral guidelines developed for the implementation of the Pollinator Plan include one specifically for gardens, "Gardening for biodiversity". This booklet is directed at garden owners of all sizes and types and its cumulative affect should be to provide more and better forage for honeybees, other insects and fauna[148]. For example, it can greatly assist pollinators if grassed areas are kept a bit on the wild side. Actions (labour saving) such as less frequent cutting of lawns or leaving a portion uncut, particularly back gardens, can encourage plants like dandelions and clovers. Clover, a low growing plant, can be encouraged if the long grass left in the lawn after the first flush of dandelions is mowed at the end of May, and then left uncut during the clover flowering season.

Since the year 2000 there has been an increasing interest in city beekeeping. Much of this has been undertaken on rooftops to remove conflict between the bees and the public at street level. It is important to have good access to the rooftop for equipment and to have hives in locations where manipulation can take place without causing a local disturbance. It is also critical that colonies have good shelter from winds. It is not unusual to see photographs of hives sited on rooftops with no or inadequate shelter. City beekeeping has developed to a stage in some cases where it is claimed the density of colonies outstrips the available forage, especially in the larger cities such as London[172]. It is also maintained that the honeybees are outcompeting wild bees for available floral sources.

**Figure 6.1a** Above (L) Gorse (*Ulex europaeus*); (R) Blackthorn (*Prunus spinosa*); Below (L) Dandelion (*Taraxacum officinale* agg.); (R) Sycamore (*Acer pseudoplatanus*). Inset are typical pollen colours (Kirk[111])

**Figure 6.1b** Above (L) Hawthorn (*Crataegus monogyna*); (R) White clover (*Trifolium repens*); Below (L) Bramble (*Rubus fruticosus* agg.); (R) Bell heather (*Erica cinerea*). Inset are typical pollen colours (Kirk[111])

**Figure 6.1c** Above (L) Ling (*Calluna vulgaris*); (R) Ivy, common (*Hedera helix*); Below (L) Oilseed rape (*Brassica napus* ssp.); (R) Apple (*Malus domestica*). Inset are typical pollen colours (Kirk[111])

**Figure 6.1d** Autumn flora, Cloonagh raised bog, County Galway showing a range of mosses and lichens incl. sphagnum moss (*Sphagnum* sp.) and *Cladonia floerkeana*

**Figure 6.1e** Cloonagh Bog, County Galway. Two colonies at the heather (ling), on a low pallet stand with a dog-leg barrier (of covered turf) providing shelter from the prevailing wind. The super on the hive in the foreground is wrapped to reduce heat loss.

**Figure 6.1f** A colony in late autumn working the ivy.

**Figure 6.1g** Above (L) *Cotoneaster horizontalis*; (R) *Hebe brachysiphon*; Below (L) Horse chestnut (*Aescules hippocastanum*); (R) Winter heath (*Erica carnea*). Inset are typical pollen colours (Kirk[111])

## 6.2 The honey harvest

There is a requirement for all *primary producers of honey* to be registered with the appropriate *National* authority. Producers may be subject to inspection to verify compliance with European Communities (Food and Food Hygiene Regulations) 2009 - SI No. 423/2009 or other appropriate *National Regulations.* Inspections will include honey hygiene, animal remedies and honey labelling (para. 5.8).

An outline of the approach to removing, extracting and storing of the honey crop will be given, only detailing areas of particular concern.

### Removing supers

It is important to have the honey ripe before extraction, namely that water content in the cells be at the appropriate level for the forage type. This is assured when the cells are sealed **(Fig. 6.2a)**. However, sometimes this may not be the case and some frames may be uncapped. In these situations, hold the frames horizontally and give them a sharp shake to test if any droplets come out. If none or only few droplets emerge it can be extracted provided the vast bulk of the other honey is sealed. If unsure, leave supers on for a day or two. A refractometer can be used to confirm the moisture content and some beekeepers will find it useful to use it to back up their own judgement **(Fig. 6.2b)**. For those selling honey there is a legal requirement to have the correct moisture level and beekeepers must be familiar with *National Regulations.* Honey with a raised moisture content will not keep, will have a tendency to ferment and should be used up first. But if being offered for sale it must comply with the regulations.

We should be mindful that honey from some forage plants, for example oilseed rape, has a tendency to granulate quickly due to its high glucose to fructose ratio. In these cases, the supers should be removed from hives before granulation occurs.

A rapid and usually reliable method to clear the bees from the supers can be achieved using the 'Canadian' type escape, although other methods such as Porter bee escapes can be used **(Fig. 6.2c)**. In areas with strong and prolonged honey flows, several supers may be 'welded' together with brace comb. It is essential that 'cracking' the supers be undertaken to avoid the honey dripping everywhere when the supers are removed for extraction. This is best done 24 hours before clearing and just before dark. The brace comb should be removed from the top and bottom of each super and put in a container. The bees will have the honey cleaned up before clearing starts. While it is a chore, it is always a good sign when the honey flows are so strong that the bees will draw brace comb to give themselves additional storage space. The author in good years with white clover and bramble in the West has generally to 'crack' supers, often extracting over 18 kg (40 lb) from each modified commercial super.

As the supers are unprotected when the bees move down, ensure that they are bee-tight, as rampant wasps or other bees may do the clearing job for you! When cleared, the supers should not be put sitting on the ground or treated in a way that can give rise to the sterile, uncontaminated product being adulterated. Small-scale beekeepers may not have a dedicated means of handling or transporting supers. Taking them off the hive and immediately into food-grade plastic sacks (preferably clear) and transporting them home in this way is well worth the effort **(Fig. 6.2d)**. They can be stored wet in the same sacks after extraction, avoiding extra work and the risk of robbing by not taking them back to the apiary again to have them 'dried out' by the bees. Attack by wax moth will also be minimised.

**Figure 6.2a** Fully capped frame of honey after clearing the bees from the super.

**Figure 6.2b** Refractometer being used to measure the moisture content of honey.

**Figure 6.2c** (L) Canadian bee escape shown from below; attached to a crown board using push pins, with second feed hole blocked off and an eke surround: (R) Porter bee escapes in crown board feed holes.

**Figure 6.2d** Supers can be put into plastic bags in the apiary for transportation. Also, returned to the bags immediately after extraction (to reduce absorption of moisture), sealed, labelled and stored wet.

## Extracting honey

It should always be remembered that honey is a super-saturated solution of sugars. To give some idea of the degree of supersaturation, a saturated solution of sucrose in water at 20°C is 66.7% (para. 5.5). On the other hand, an average composition of honey would be 18% water and 79% sugar: 40% fructose, 35% glucose and 4% other sugars[105]. Hence honey has a great propensity for water (hygroscopic) and when out of the sealed cell and in contact with humid air will absorb water to bring it back toward its *stable* sugar saturation level. We should be mindful of this property when working with honey; extract in a dry environment, keep honey in containers well sealed with small air space and seal up wet supers immediately after extraction. The missing 3% in the composition of honey above gives each honey its unique characteristic and will be discussed later (para. 6.3).

Honey should be extracted as soon as possible after being removed from the hive. It helps if the supers can be held in a dedicated room that is dry and warm to at least retain some of the heat that was in the supers when they were taken off the hive. Honey is a food and clean and hygienic practices must be used.

This room may also be used for extraction, but in most cases the small-scale beekeeper will be extracting in the domestic kitchen. *National Regulations* will apply to this operation with two water sources, one for hands and equipment and the other for washing the floor.

Frames can be placed on a rebated lath resting on top of a bucket **(Fig. 6.2e)**. Cell cappings are usually removed with a long, slim serrated knife using an upwards action. An uncapping fork is useful to remove cappings that are in a 'dip' on the face of the comb. Cappings can fall into the bucket and the finished uncapped frame inserted into an extractor. Where there are variations in the honey-fill of the frames, it is important as far as possible to balance the frames by weight across the diameter (opposites) of the extractor. If this imbalance is extreme, the nett centrifugal force across the diameter can be large, vibrating the extractor, damaging comb and reducing the life of the extractor. For that reason, the rate of speed build-up should be low to get a feel for the degree of imbalance. However, this imbalance may not be just a case of waiting for the cells to be emptied to achieve balance. Those who have colonies in areas with nectars, such as oilseed rape that granulate quickly, may have a persistant imbalance problem. In these cases, the extractor speed needs to be closely watched.

Extracted supers can be sealed in the sack, labelled and stored **(Fig. 6.2d)**. Of course, some beekeepers prefer to store dry, and in that case the supers will need to be taken back to the apiary, preferably put back on the same colonies, and then taken back again for storage. There will be cases where there is a lot of granulated honey still left after extraction and these can be fully immersed in a water bath **(Fig. 6.2f)**. Heating the water will speed up the process, but great care must be taken to ensure that the temperature does not go above 30°C as comb strength will be weakened and comb damage will result when shaking out the water. A day later the supers can be taken off and the water shaken out, and left in an airy location, indoor or outdoors until dry. A second immersion in water may be needed.

## Bottling/storage

The extracted honey can be filtered straight into a settling tank and after 24 hours the honey bottled. Honey that is prone to granulate quickly will after a short time in the jar set hard and also 'frost', indicated by shrinkage away from glass and have a white, cloudy appearance. Ripe honey, to be stored longer-term, can be put in buckets after extraction, labelled and stored at a low temperature, less than 10°C preferred, to deter fermentation.

This stored honey can later be brought back to a clear or runny honey state, as required, using a thermostatically controlled heating cupboard or a water bath **(Fig. 6.2g)**. A water bath gives good thermal contact and will bring the honey up to the required temperature quickly, and without the need for an excessive temperature. Remember to put lids back on the buckets when heating. Honey must not be overheated as it will raise the hydroxymethylfurfural or HMF levels, a cumulative measure of age and temperature. Another measure of overheating is the enzyme diastase. There are *National Regulations* for these two measures. There are also regulations covering record keeping for the honey crop as well as labelling of jars with appropriate Lot Numbers to allow traceability.

## Soft-set (creamed) honey

Most beekeepers who have colonies in country areas will have encountered honey that granulates fast, such as oilseed rape in the brassica family. Small-scale beekeepers will generally have more difficulty than larger operations in achieving a stable honey without excessive heat being used. Soft-set (creamed) honey is something that offers an attractive alternative. It is easily spread with an agreeable texture. Achieving a good soft set-honey will require some 'trial and error' over time as there is a bit of an art involved that needs time to perfect.

Where the honey has a coarse grain:
- take the bucket of granulated honey (or run honey that you want to cream) and scrape the scum off the top.
- heat the honey slowly until liquid/clear of granulation **(Fig. 6.2g)** and filter into a settling tank or large bucket with tap. Allow it to cool to 20/25°C.
- the honey needs to be seeded with about 10 per cent of a known source of soft-set honey with a good texture, preferably one that has been known to be stable. This soft-set honey should be gently heated until it flows as a thin paste, but one must ensure that crystals do not dissolve.
- add the soft-set seed to the honey in the bottling tank or bucket and mix well with slow stirring action that minimises the introduction of air.
- Allow it to settle overnight in a warm room and the honey can then be bottled and stored at a lower temperature, 10 to 15°C.
  Note:
  If the honey in the bucket has a fine grain it can be creamed without a seed. The granulated honey can be heated slowly to about 25/30°C until it has softened and flows as a paste. It should then be stirred, allowed to settle overnight, bottled and stored as described above where a seed has been used **(Fig. 6.2h)**.

**Figure 6.2e** Uncapping (decapping) a frame of honey with upward action and tilted frame, allowing the cappings to drop into a bucket below. Inset, an uncapping fork.

**Figure 6.2f** A stainless-steel water bath with two supers containing granulated honey in various stages that have been uncapped and immersed in water to dissolve the granulation.

**Figure 6.2g** Bucket of stored honey being brought back to a runny or clear state in a thermostatically controlled water bath.

**Figure 6.2h** A jar of creamed honey. It can be produced directly from a fine-grained honey or by adding a fine-grained seed to a coarse-grained honey.

## 6.3 Honey, some perspectives

### Honey, the health food

Honey is produced by the bees for their own consumption, and along with pollen provides all their needs for energy, growth and development. In the hive, honey with its high acidity and osmotic pressure is a hostile medium to bacteria.

Honey is used as a healthy food supplement to the human diet, but increasingly it is being recognised for those medicinal properties demonstrated in the brood nest. Research in Ireland published in 2019 gave a detailed insight into the beneficial properties of honey[110]. The total phenolic content (TPC) of honey is linked to its antioxidant and antimicrobial properties. The study looked principally at the TPC of honey over a range of floral types, forage locations and made international comparisons. It showed that Irish ling (heather) honey had a similar TPC to Manuka honey. Oilseed rape had the lowest level for Irish unifloral honeys, while ivy honey was about half that of heather. Urban multifloral honeys had about a one third higher TPC value than that from rural areas. It was felt that the diversity of floral plants was the important factor. The physiochemical characteristics of Irish heather and Manuka were also similar. These findings contrast with those from a Polish study published in 2020, which showed that Manuka honey had over twice the TPC content of heather[109]. In fact, the TPC values for Irish heather were a remarkable three times higher than for heather *Calluna vulgaris* measured in recent Polish papers[109,161]. Part of the variation in the comparison with Manuka can also be explained by the measured reference TPC values for Manuka honeys used in the Irish study being lower than those in international studies.

While there was a reasonably large spread in some of the measured values reported, and it was suggested that further study of Irish heather honey should be undertaken, there are strong indications of the important health benefits of Irish honeys.

### Colour, flavour and aroma

Every honey has its own special characteristics. They are all different, all largely determined by the variety of plants on which the bees have foraged. The wide range of colours, flavours and aromas adds a great interest to the honey harvest, and for none more so than the beginner-beekeeper. The beginner will always remember the first crop after taking up beekeeping, even down to the number of jars. The author's first crop of eight jars of very dark honey was harvested in late July after a warmer than normal summer. Two jars were taken to the local honey show and the judges were confounded. It transpired to be honeydew with a lot of ash, was slow to granulate and the honey continued to take a prize in subsequent years.

With monofloral crops such as oilseed rape, one is likely to get a fairly pure sample of one particular blossom. However, generally, in country (rural) areas the honey is determined by the nectars from a number of plants. In these areas it is still possible to get a reasonably pure honey of one particular plant if there is a strong flow from that plant with no other major source in blossom. An example of this once in a decade event in County Galway relates to sycamore and hawthorn, where weather conditions, including a late frost, have affected the blossoming of one or other plant leaving the other dominant **(Fig. 6.3a)**. Similarly, for white clover where the flow from the old-lea clover can reach up to 4kg per day swamping the later blossoming bramble (blackberry) **(Fig. 6.3a)**. Ling (heather) is much more predictable where a strong flow can usually overwhelm those from other minor plants in bloom on the periphery such as knapweed **(Fig. 6.3a)**.

Cut comb and soft-set honey will also add variety to the honey range. Beekeepers, and in particular beginners and those with one or two colonies, can still add variety to their honey crop by taking it off in late spring as well as late summer. In urban areas the honey is a cocktail of many forage plants and tends to be consistent from year to year. Honey from the second half of the season is normally darker in colour, probably resulting from the different seasonal blossom and higher temperature. There is, however, a strong correlation between colour and TPC content indicating that the later season honey in urban areas may have greater health properties[110,161]. Examples of the contrasting honey from an urban town are shown **(Fig. 6.3b)**. It is also interesting to compare the three clear honeys from the rural locations **(Fig. 6.3a)**, with the two much darker honeys from the urban location **Fig. 6.3b)**. This is consistent with the findings in the Irish study that urban multi-floral honey is darker and has a higher TCP than honey from rural locations[110].

## Showing / demonstrating honey

Beekeepers and beekeeping associations have a role to play in informing the public on bees and beekeeping, as well as biodiversity. Local annual events are a good outlet to involve the public. An observation hive can attract a lot of attention and be a good way to give an insight into the insect world and the importance of pollinators **(Fig. 6.3c)**. It is stressful for a colony to be in an enclosed observation hive for more than 24 hours. Some associations may be in a position to have a permanent observation hive in a public building where the small colony can reside for all or a good part of the year. The bees will be free to fly to the outside (via a tube), but such an arrangement needs to be managed to ensure that the colony does not become congested. A variation on this is where a full production colony is housed in a hive that has visual access to one of the sides. This requires a section to be removed from one side and replaced by a perspex insert (**Fig. 6.3d**). A removable wooden side can be placed on the outside and fixed with thumb screws. This set-up can be completed by fixing a black plastic covering to act as a canopy over the observer to made it light-proof, a bit like the photographers of old. An arrangement such as this can be used in an association's beginner or demonstration apiary. Sitting on a stool, draped by a canopy reaching to the ground, along with a small torch allows a different perspective of the honeybee colony (as honeybees do not recognise the colour red, a red covering over the torch will allow the bees to be seen, yet unaffected). Observing the bees working away on the frames, undisturbed in the dark, can give beekeepers (and particularly beginners) a different insight into a honeybee colony and a new respect for bees.

Honey shows and honey tasting events can also stimulate interest and being able to taste a variety of honey types will add further to the experience (**Fig 6.3e**). Not all beekeeping associations will be in a position to mount a full honey show event. However, an effective way to showcase honey is to have a honey-tasting night and invite the public. The jars brought in can be lined up, light to dark, and members given one minute each to explain the provenance of their honey followed by a tasting. It is a good way to get member-involvement, and members who may not have spoken for the rest of the year will always have something to say about their honey.

## Honey for 'retired' beekeepers

When beekeepers of a long standing are forced to give up the craft through age or infirmity, the change can be both abrupt and traumatic. There is a lifestyle change and the jars of honey that were always there are no longer. These beekeepers are likely to have promoted beekeeping in their neighbourhood by assisting their fellow beekeepers over the years or by contributing to the running of their local beekeeping association.

Beekeeping friends may provide honey informally, but in their last few years of beekeeping their contact with fellow beekeepers can be reduced as their profile declines. A formal arrangement whereby Association members put a jar of honey each year into a honey pool is likely to be the more reliable way of ensuring a supply of honey that can be made up into gift packs (**Fig. 6.3f**). This should help to ease the transition from no bees and no honey. The partners of beekeepers, who equally were used to a supply of honey, may also be affected when the beekeeper passes on, and the supply of honey should equally go to them. At that stage of life, small things can make a big difference.

Figure 6.3a A diverse flora will add variety; colour, taste, aroma to the honey. Jars of honey, above (L) sycamore, light amber; (R) hawthorn, medium amber; below (L) white clover, light straw and (R) ling, all from areas where a dominant forage species was present.

Figure 6.3b Urban honey jars from the same colony; (L) early season and (R) late season. Taking honey off mid-season will add variety; colour, aroma and taste.

**Figure 6.3c** Observation hive: below; a brood frame of pollen, honey stores, brood (open and sealed), worker/drone bees and marked queen. Above a super frame with both sealed and open honey cells.

**Figure 6.3d** A Modified Commercial hive with a section of the side removed, exposing the end bars of the Hoffman frames. The black plastic above is the upper part of a canopy that is pulled over the observer sitting on a stool with a small torch, allowing the colony to be observed working in the dark between frames.

**Figure 6.3e** A display rack at a public event showing a seasonal range of bee forage along with corresponding honey pot tasters; oilseed rape, hawthorn, clover, bell heather, ling (heather) and an urban honey.

**Figure 6.3f** A gift pack of honey to be given to retired beekeepers of long standing.

# Glossary

The items listed in this glossary are referenced to the relevant paragraph(s) in the book [ ] and where more than one paragraph, ranked in order of importance

**Abdomen**
The posterior or third segment of an insect; in the honeybee it contains many vital organs, as well as circulatory, respiratory, reproductive, digestive and excretory systems, [2.2]

**Acari**
Order within the animal Kingdom that includes tracheal, varroa and tropilaelaps mites [4.0]

**Acarine**
Disease of adult bees caused by a microparasite, *Acarapis woodi.* Condition also referred to as *acariosis,* tracheal mites and 'Isle of Wight disease' [4.5,2.1]

**Acute bee paralysis virus**
Viral disease that is largely vectored by varroa mites [4.12]

**Alimentary system**
Facilitates digestion of food, and includes pharynx (mouth), honey crop/stomach, ventriculus (stomach), small intestine and rectum [2.2]

**American foul brood (AFB)**
A brood disease caused by the spore forming bacteria *Paenibacillus larvae.* Very virulent, it is a notifiable disease in most countries. [4.1]

**Antenna**
One of two long, segmented projections in the head of a honeybee, with sensory organs; mainly of smell, taste and touch [2.2]

**Anther**
Male part of a flower containing the pollen [2.4]

**Apiary**
A site where honeybee colonies are located for beekeeping purposes [3.1]

**Apiculture**
The science and art of beekeeping

**Apis**
Genus in which the honeybee is classified [2.1]

**Apis mellifera (A.m.)**
The common or European honeybee species now found throughout the beekeeping world. Four main subspecies in Europe; North European dark bee *A. m. mellifera*, the Italian *A. m. ligustica*, the Carniolan *A. m. carnica* and the Caucasian *A. m. caucasica* [2.1]

**Arthopoda**
Phylum, subdivision of invertebrates, in the animal Kingdom with segmented bodies. It includes insects (Insecta) and mites (Arachnida/Acari) [2.2]

**Artificial swarm**
Control measure to prevent swarms and used to raise queens or new colonies [5.2]

**Bacterium**
A large group of microorganisms of which a few are parasites of the honeybee [4.0,4.1,4.2]

**Bee escape/clearer board**
A one-way exit used to move bees out of the honey super. Porter and Canadian style bee escapes are examples [6.2]

**Bee bread**
A mixture of nectar and pollen that is stored in the hive and fed to young worker and drone larvae [2.2]

**Bee space**
A gap of ~ 8 mm allowed between hive parts that is small enough to discourage the bees from building brace comb and large enough not to be filled with propolis [3.2,5.3b]

**Beeswax**
A complex mixture of organic compounds secreted by the bee's wax glands. Its melting point is 64°C [2.2]

**Brace comb**
Small, irregular pieces of wax comb joining two normal combs or filling up gaps [6.2]

**Brood**
Immature colony members, eggs, larvae and pupae. Before capping it is referred to as *unsealed* brood and after capping *sealed* brood. Capped worker brood and stored honey/pollen cells are flat while drone cappings are domed [2.2, 5.2]

**Brood box**
Part of the hive with wax comb where the queen and brood reside along with stored honey and pollen [3.2, 3.3, 5.2]

**Brood food**
Food derived from the hypopharyngeal and mandibular glands of workers and fed to larvae of workers and drones [2.2, 4.2,4.4]

**Capping**
The material covering a sealed cell. Wax only for honey cells and wax plus other porous material for brood cells [2.2, 5.5]

**Cast**
A second or subsequent swarm, with an unmated queen, issuing from a colony that has already swarmed [5.2]

**Caste**
One of three functionally different honeybees or castes within a colony; the two female castes, worker and queen and the male drone [2.2]

**CDB Hive**
A double-walled hive developed by the Congested Districts Board (CDB) which was set up in 1891 to improve living standards in rural Ireland. It is a double-walled gable type hive, but it has largely been replaced by the National or Modified Commercial hives [3.2]

**Chalk brood**
A brood disease caused by the fungus *Ascosphaera apis* [4.3]

**Chronic bee paralysis virus**
Viral disease that infects all stages of the honeybee: eggs, larvae and adults [4.9]

**Circulatory system**
Facilitates the circulation of haemolymph (blood) and includes the heart, dorsal/ventral diaphragms, blood and aorta [2.2]

**Cleansing flight**
Bee flight to allow the bees to empty their bowels of the accumulated waste products during winter [2.2]

**Clearer board**
A piece of equipment used to clear bees out of the supers at honey extraction time. May be a crown board fitted with Porter or Canadian bee escapes [6.2]

**Clipping a queen**
Trimming a portion of a queen's wing to restrict her flying ability, used as part of a swarm-control method [5.3c]

**Clustering**
Refers to the honeybees' trait of clinging together in a tight mass. It applies to swarms and to the winter cluster, and is effective in reducing heat loss [2.3, 5.2]

**Colony collapse disorder (CCD)**
A term used to describe a condition in North America during the early 2000s where bees would abscond the nest leaving weak or dead bees and a comparatively large area of brood [4.15]

**Commercial hive**
A single-walled 'square' hive, sometimes referred to as a Modified Commercial. It has a comb area about 50 per cent larger than a CDB/WBC and one third larger than the National hive. Its use is largely confined to Britain and Ireland [3.2, 3.3]

**Corbicula**
The pollen basket on the outside of the two hind legs of worker bees for carrying the pollen load. Also for carrying propolis back to hive [2.2]

**Cracking (the supers)**
Breaking (cracking) the supers apart, a day before putting on clearer boards, to remove brace comb and allow the bees to clear up honey. Generally, only required after very strong honey flows [6.2]

**Crown board**
A wooden cover placed over the brood box or super [3.2, 5.2]

**Cuticle**
The tough, flexible outer covering of the bee that provides physical protection. Its epicuticle covering gives a waterproof seal [2.2]

**Dances**
The means used by honeybees to communicate the location of forage (food) or a suitable home for a swarm. The round dance indicates that there is food nearby, while the waggle (figure-eight) conveys both direction and distance [2.2]

**Deformed wing virus**
Viral disease that infects all castes of the honeybee: workers, drones and queen. It is vectored by *Varroa* mites [4.11]

**Diastase**
An enzyme that breaks down starch [6.2]

**Diseases (honeybees)**
A group of micro and macro-organisms that are parasites of the honeybee [4.0]

**Drifting**
The entering of other hives by individual bees that fail to find their own hives after a flight [4.1,2,5,6]

**Drone**
A male honeybee [2.2]

**Drone congregation area**
A geographical location in mid-air where drones gather to await the arrival of virgin queens [2.2, 5.1, 5.2]

**Drone-laying queen**
Queen due to age, imperfect or no mating and whose eggs will only be drones or progressively be drones [4.14b]

**Dummy board**
A moveable partition of the same overall size as the brood frames, that is used to replace one of the brood frames and thereby reduces the danger of rolling and squashing bees when the first frame is being removed [3.2]

**Dysentery**
Condition where bees leave excrement on or in the hive [4.13d]

**Ectoparasite**
A parasite living on the outside of the host [4.0]

**Endoparasite**
A parasite living within the host [4.0]

**Eukaryote**
Single and multi-celled organisms that contain their genetic material within a membrane-bound nucleus. Examples are fungi/microsporidia [4.0]

**European foul brood (EFB)**
A brood disease caused by the bacteria *Melissococcus plutonius*. It is a notifiable disease in most countries [4.2]

**Excretory system**
The excretory system includes Malpighian tubules, small intestine, rectum and anus. The tubules absorb waste matter from the haemolymph (blood) and passes it into the small intestine. The

rectum stores the faeces and is capable of expanding to fill a large part of the abdomen during winter between cleansing flights [2.2]

**Fat bodies**
A tissue found throughout the body of a larva and adult bee that acts as a food store [4.6,6.1]

**Feed hole**
Aperture in the crown board through which bees can be fed [5.5]

**Feral colonies**
Wild colonies that are not under the influence of a beekeeper. Traditionally in tree cavities [3.2]

**Fertilisation**
Process by which the female gamete and male gamete come together to form new life [2.2,2.4]

**Foragers**
Bees whose task is to collect nectar, water, pollen and propolis. Usually the older bees [2.2,2.4]

**Foundation**
Thin sheet of beeswax embossed with the pattern of the base of cells [5.2]

**Frame configuration ('warm'/'cold')**
'Warm' way, a method of arranging frames so that the top bars are parallel to the entrance. The 'cold' way is when the top bars are at right angles [3.3]

**Frosting**
A white, cloudy appearance in a jar of honey due to rapid granulation and shrinkage away from the glass [6.2]

**Fructose**
One of the two main sugars in honey [6.2]

**Gamete**
A cell containing only half the DNA contained in a normal cell and are the reproductive cells of egg and sperm [2.2]

**Genus**
Group of closely related organisms classified between Family and Species. Honeybees are classified as *Apis* [2.1]

**Glucose**
One of the two main sugars present in honey, and its raised concentration increases rate of granulation [6.2]

**Grafting**
Moving very young, just hatched, larvae from a brood cell to an artificially constructed queen cell in the process of mass queen rearing [5.1]

**Grooming**
The removal of foreign matter from its body (autogrooming) or the body of a nestmate (allogrooming). Includes removal of pollen and also relates to parasite removal [4.0]

**Guard bee**
A nurse bee (2-3 weeks old) whose duty is to remain at the hive entrance and prevent worker bees from another colony or wasps etc entering [2.2]

**Haemolymph**
The blood (liquid) in the body of the honeybee. The whole of the body is filled like a sump (it does not have veins), it delivers nutrients (does not deliver oxygen to cells) and is under pressure to maintain body shape [2.2]

**Hefting**
Used to make an estimate of hive weight and hence weight of stored honey. Crude measure, and can be misleading in inexperienced hands [5.5]

**Honey crop**
The crop/stomach is an expandable portion of the alimentary system that holds water, and also nectar during foraging and regurgitated as part of honey processing [2.2]

**Honeydew**
Honey produced from secretions of sap-sucking insects or extrafloral nectarines [6.3]

**Host**
Organism in which the parasite/pathogen actively replicates [4.0]

**Hydroxymethylfurfuraldehyde (HMF)**
A substance produced by the breakdown of sugars largely due to the cumulative effect of age and temperature. Honey for sale must meet standards in *National Regulations* [6.2]

**Hymenoptera**
Order of insects with two pairs of membranous wings that includes bees, wasps and ants, [2.2]

**Hypopharyngeal gland**
One of a pair of glands in the head of the honeybee, that produces an additive to brood food and enzymes for nectar ripening [2.2,4.4]

**Integrated pest management (IPM)**
A system to reduce pest levels using timely applications of different approaches. Used in recent years to contain *Varroa* mites [3.2]

**Isle of Wight disease**
Honeybee disease condition first identified on the Isle of Wight (off southern England) in 1904/05, and also referred to as acarine, tracheal mites, *Acarapis woodi*, and *acariosis* [4.5,2.1]

**Langstroth hive**
A single-walled hive most widely used worldwide. Its brood area size lies between the National and Modified Commercial hives, the latter two being the most popular hives in Britain and Ireland [2.1, 3.2]

**Larva**
A rapid development stage between egg and pupa when brood is fed [2.2]

**Laying workers**
Worker bees that become capable of laying a few eggs, in the absence of the queen. Any bees that result with be stunted drones [4.14c]

**Locally-adapted colonies**
Colonies that have been raised within a regime that restricts the introduction of bees from outside. In effect the gene pool in the area is contained, and adaption (selection) of the indigenous bee takes place. Non-treatment of colonies may also be a feature, see 'sustainable colonies' below [5.1,5.2]

**Mandibles**
Two mouth mandibles to chew wax and intruders / parasites (e.g. varroa mites) [2.2]

**Manipulations**
Beekeeping operations that involve removal or exchange of comb in a colony [5.3]

**Mating flight**
A queen's flight to a drone congregation area to be mated by drones. During a few mating flights the virgin queens can mate more than twenty times [2.2]

**Mating nuc**
A small colony into which ripe queen cells are placed. Virgin queens will fly and mate from here [5.2]

**Microsporidia**
Group of intercellular organisms affecting the intestines of their host [4.0, 4.4]

**Mites**
Group of highly specialized animals and classified within the order *Acari* [4.0, 4.5,4.6,4.7]

**Monofloral honey**
Honey derived from one flower type [6.3]

**Moulting (ecdysis)**
Process of getting rid of old cuticle and growth of the new one during larval and pupal stages [2.2]

**Nasanov gland**
Gland located at the rear of the abdomen which secretes a pheromone that attracts other bees [2.2]

**National hive**
Single-walled 'square' hive has a comb area not much larger than a CDB/WBC hive and about 30 per cent smaller than a Commercial hive. Its use is largely confined to Britain and Ireland [3.2,3.3]

**Nectar**
Sugary liquid secreted, usually by flowers, and collected by bees to produce honey [2.4, 6.1]

**Nosema**
Two diseases of adult bees caused by the microsporidia, *Nosema apis* and *Nosema ceranae,* that have recently been reclassified as specialized parasitic fungi. The parasites infect the lining of the ventriculus (gut) [4.4]

**Notifiable Disease**
Disease stipulated *by National Regulations* that must be notified to the governing authority, for example, American foul brood [4.1,4.2,4.7,4.8]

**Nucleus box**
A small hive that can contain a small colony for queen mating or start up colony before expanding into a full hive [5.1,5.2,3.3]

**Nurse bees**
Workers that undertake the duties in the brood nest, generally during their first ~3 weeks **[2.2]**

**Parasites**
Living organism that finds a suitable host (honeybee) in or on which to undertake its lifecycle and generally obtains nourishment at the host's expense [4.0]

**Parasite load/intensity**
The average number of parasites present in a diseased population and is expressed as the number of mites, spores, particles per bee [4.0]

**Parent colony**
Original colony in which the bees resided. In a swarming situation, the colony from which the swarm emerged [5.2]

**Pathogen**
An organism that can cause disease, can be referred to as a microparasite [4.0]

**Pheremones**
Chemical substance released externally by one animal to stimulate a response in another of the same species [2.2,2.3]

**Phoretic stage**
Migratory or questing stage of a lifecycle [4.6]

**Piping**
Hollow sound made by a queen. It can indicate that a virgin queen has emerged [5.3]

**Polarised light**
Light waves that are orientated in one plane. Feature used by bees in applying the directional input from dances [2.2]

**Polished cells**
Cells that are polished is an indication that they are being prepared to receive an egg. This also applies in the case of a queen cell [4.14, 5.2]

**Pollen**
Substance produced by the male part of a flower, and collected by the worker bees which provides the protein in the brood and bee diet [2.4]

**Pollen basket (Corbicula)**
Spiny/hairy structure on hind legs of worker bee used to carry pollen back to the colony. Not in drones or queens.

**Pollination**
Process of transfer of pollen grains from a ripe anther (male part) to a receptive stigma (female part) of a flower of the same species. Once pollinated, a tube grows through the stigma to enable

fertilization of the ovary below [2.4]

**Polyandry**
The multiple mating of honeybees. This gives the honeybee a greater spread of attributes including an increased tolerance to disease [4.0]

**Prevalence**
Proportion of the total colony population that has a disease, usually expressed as a percentage [4.0]

**Prokaryotes**
Group of single-celled organisms whose genetic material is not contained within a nuclear membrane but distributed throughout the cell. Includes bacteria [4.0]

**Propolis**
Sticky substance collected by bees from resinous buds of plants and carried back to the colony. Sometimes referred to as 'bee glue', it has antiseptic properties [4.0]

**Pupa**
Stage between lava and adult bee during which the final adult state is formed. Cell is capped with wax and other material to make it porous [2.2]

**Queen**
The reproductive female in the colony that develops differently from the larva of a worker by being fed only on royal jelly. She has the capability to lay a fertilized egg (for worker or queen) or unfertilized egg (drone) [2.2]

**Queen cell**
An elongated wax cell hanging vertically from a comb, and built for the development of a queen bee from egg/larva/pupa to emerging adult [2.2,5.2]

**Queen cup**
Cell hanging vertically from a comb, resembling an acorn cup and sometimes referred to as a play cell. At start of the swarming season, they are distributed around the brood nest, like queen-cell foundation, to be used when required [2.2, 5.2]

**Queen excluder**
A grid of wire or perforated metal placed over the brood box that prevents the queen from entering the super but admitting workers [3.3, 5.2]

**Quarantine pests**
Pests of potential economic importance that are being officially controlled [4.7,4.8]

**Queenright**
A colony with a living, laying queen [2.2]

**Queen substance**
Group of pheromones secreted by the queen that attracts and indicates queen's presence to worker bees. Its presence, in sufficient quantity, is considered to regulate swarming [2.2, 5.2]

**Refractometer**
Device that measures the water content in honey by using its refractive index [6.2]

**Respiratory System**
Facilitates the exchange of oxygen and carbon dioxide and includes spiracles (vents in thorax and abdomen) and a series of tubes (tracheae, tracheal sacs and tracheoles) [2.2]

**Ripe queen cell**
Queen cell that is close to hatching. Worker bees will remove wax from tip to expose fibrous cocoon [5.2]

**Royal jelly**
A substance secreted from the mandibular and hypopharyngeal glands of workers and fed to queen larvae [2.2]

**Sacbrood virus**
Disease long associated with honeybees and caused by the *Sacbrood virus* that infects the bee larvae [4.10]

**Scout bees**
Worker bees that independently look for new sources of forage, or new nest sites [2.2]

**Signs/Symptoms**
Physical or behavioural changes caused by the host's adverse reaction to the presence of a disease [4.0]

**Skep**
Traditional hemispherical hive usually made from straw with naturally drawn comb. Replaced from mid-1800s by moveable frame hives [3.2]

**Small hive beetles**
Disease caused by the beetle *Aethina tumida* that feeds on honey, pollen and brood. They are harmful parasites of honeybees and to date their presence in Europe has been confined to southern Italy. It is a quarantine species, and its presence must be notified to the national authority [4.8]

**Spermatheca**
An organ within the queen's abdomen where she stores the sperm received during mating [2.2]

**Spiracles**
Apertures along sides of thorax and abdomen that are openings to breathing tubes or trachea [2.2, 4.5]

**Stigma**
The female part of a flower receptive to the pollen (male part) [2.4]

**Superorganism**
A large group of individuals working together for a common cause and taking on the appearance of a single entity [2.3]

**Supers**
Upper sections of a hive where the bees store the honey [3.3]

**Supersedure**
The process by which the colony can replace an old queen by a new one without swarming [4.4,5.2]

**Sustainable colonies**
Colonies that react, as they would in nature, to changes that are detrimental to their health or development, without the need for any beekeeping treatments. Colonies are also locally-adapted, see 'locally-adapted colonies' above [5.1,5.2]

**Swarm control**
Beekeeping method used, after signs of swarming by the colony, to stop a swarm leaving a hive [5.2]

**Swarming**
The natural method of reproduction in honeybees whereby the queen and about half of the parent-colony workers (swarm) leave and settle on a site nearby [2.3, 5.2]

**Swarm prevention**
Beekeeping methods taken to prevent or delay the colony from making swarm preparations [5.2]

**Thorax**
The part of the bee between the head and the abdomen. It is the 'chassis' to which the legs and wings are attached, and contains the strong muscles for flight and colony heating [2.2]

**Tracheae**
The internal breathing tubes of the bee that start at the spiracles on the body surface [2.2]

**Tracheal mites**
Disease of adult bees caused by a tracheal mite *Acarapis woodi.* Condition also referred to as *acariosis,* acarine and 'Isle of Wight disease' [4.5,2.1]

**Trophallaxis**
The transfer of liquid substances (food) from one honeybee to another [5.2]

**Tropilaelaps mites**
Disease caused by three species of mites that feed on bee brood: *T. clarease, T. koenigerum and T. mercedesae.* To date have not been found in Europe and

are therefore quarantine pests. Their presence must be notified to the national authority [4.7]

**Uncapping (decapping)**
Removing the wax cappings on sealed honey cells to enable honey to be extracted [6.2]

**Varroa**
Disease of brood and adult bees caused by the external varroa mite, *Varroa destructor* [4.6]

**Vector**
An organism with parasites that enables the parasites to be transmitted to new hosts [4.0]

**Ventriculus**
Stomach (midgut) where digestion takes place [2.2,4.4]

**Virulence**
The virulence of a parasite is its capacity to harm its host [4.0]

**Wax glands**
Four pairs of glands on the underside of the worker bee's abdomen [2.2]

**WBC hive**
A double-walled gable type hive designed in late 19th century in Britain by W.C. Carr and similar to the CDB in Ireland. The hive has largely been replaced by the National or Modified Commercial hives [3.2]

**Winter bees**
Worker bees that are raised in the autumn, with developed fat bodies, that the colony depends on to see it through the winter and raise the young bees in the early spring. They must live many months rather than weeks, as is the case at other times of the year [2.2, 5.5]

**Worker bee**
One of the females in a colony that are not re-productive [2.2]. These bees carry out all the routine work in the colony. In the absence of a queen may be capable of laying eggs, and in this case are called 'laying workers' [4.14c]

# Bibliography


1. Adam Bro. (1968) "Isle of Wight" or acarine disease: Its historical and practical aspects, *Bee World* 49, 6-18.

2. Adam Bro. (1975) Beekeeping at Buckfast Abbey, Northern Bee Books, Hebden Bridge, UK.

3. Akratanakul P., Burgett M. (1975) *Varroa jacobsoni*: a prospective pest of honeybees in many parts of the world, *Bee World* 56, 119-121.

4. Alaux C., Folschweiller M., McDonnell C., Beslay D., et al. (2011) Pathological effects of the microsporidium *Nosema ceranae* on honey bee queen physiology (*Apis mellifera), J. Invertebr. Pathol.* 106, 380-385.

5. Alan A.J., Carvajal M.A.,Vergara P.M. (2020) Giants are coming? Predicting the potential spread and impact of the giant Asian hornet (*Vespa mandarinia*, Hymenoptera:Vespidae) in the USA, *Pest Management Science*, doi 10.1002/ps.6063

6. Allanby M. (2009) Oxford Dictionary of Zoology, Oxford University Press, Oxford.

7. Allen M., Ball B.V. (1996) The incidence and world distribution of honey bee viruses, *Bee World* 77, 141-162.

8. Anderson D.L., Gibbs A.J. (1988) Inapparent virus infections and their interactions in pupae of honey bee (*Apis mellifera* Linnaeus) in Australia, *J. Gen. Virol.* 69, 1617-1625.

9. Anderson D.L., Morgan M.J. (2007) Genetic and morphological variation of bee-parasitic *Tropilaelaps* mites (Acari: Laelapidae): new and redefined species, *Exp. Appl. Acarol.* 43, 1-24.

10. Antúnez K., Harriet J., Gende L., Maggi M., et al. (2008) Efficacy of natural propolis extract in the control of American foulbrood, *Vet. Microb.* 131, 324-331.

11. Bailey L. (1957) The isolation and cultural characteristics of "*Streptococcus pluton*" (*Bacillus pluton* White) and further observations on *Bacterium euridice* White, *J. Gen. Microb.* 17: 39-48.

12. Bailey L. (1958) The epidemiology of the infestation of the honeybee, *Apis mellifera* L. by the mite *Acarapis woodi* Rennie and the mortality of infested bees, *Parasitology* 48: 493-506.

13. Bailey L. (1960) The epizootiology of European foulbrood of the larval honey bee *Apis mellifera* Linnaeus, *J. Insect Pathol.* 2, 67-83.

14. Bailey L. (1961) The natural incidence of *Acarapis woodi* (Rennie) and the winter mortality of honey-bee colonies, *Bee World* 42, 96-100.

15. Bailey, L. (1964) The "Isle of Wight disease": The origin and significance of the myth, *Bee World* 45, 32-37, 18.

16. Bailey L. (1965) Paralysis of the honey bee *Apis mellifera* Linnaeus, *J. Invertebr. Pathol.* 7, 132-140.

17. Bailey L. (1965) The occurrence of *chronic* and *acute bee paralysis viruses* in bees outside Britain, *J. Invertebr. Pathol.* 7, 167-169. *J. Inverteb. Pathol.*

18. Bailey L. (1974) An unusual type of Streptococcus pluton from the Eastern Hive bee, 23, 246-247.

19. Bailey L. (1976) Viruses attacking the honey bee, in: Lauffer, M. A., Bang, F.B., Maramorosch, K., Smith, K.M. (Eds.) *Advances in virus research*, vol. 20, Academic Press, New York, pp.271-304.

20. Bailey L. (1981) Honey bee pathology, Academic Press, London, UK.

21. Bailey L. (1983) *Melissococcus pluton*, the cause of European foulbrood of honey bees (*Apis* spp.), *J. Appl. Bacteriol.* 55, 65-69.

22. Bailey L., Ball B. V. (1991) Honey Bee Pathology, Academic Press, London.

23. Bailey L., Ball B.V. (1997) Honey Bee Pathology, Academic Press, London.

24. Bailey L., Collins M.D. (1982) Reclassification of "Streptococcus pluton" (White) in a new genus *Melissococcus*, as *Melissococcus pluton* nom. rev.; comb. nov. *J. Appl. Bacteriol.* 53, 215-217.

25. Bailey L., Gibbs A. J., Woods R. D. (1963) Two viruses from adult honey bees (Apis mellifera Linnaeus), *Virology* 21, 390-395.

26. Ball B.V., Allen M.F. (1988) The prevalence of pathogens in honey bee (*Apis mellifera*) colonies infested with the parasitic mite *Varroa jacobsoni, Ann. Appl. Biol.* 113, 237- 244.

27. Ball B.V., Overton H.A., Buck K.W., Bailey L., et al. (1985) Relationships between the multiplication of chronic bee- paralysis virus and its associate particle, *J. Gen. Virol.* 66, 1423- 1429.

28. Bastos E.M.A.F., Simone M., Jorge D.M., Soares A.E.E., et al. (2008) In vitro study of the antimicrobial activity of Brazilian and Minnesota, USA propolis against *Paenibacillus larvae, J. Invertebr. Pathol.* 97, 273-281.

29. Beaurepaire A., Piot N., Doublet V., Antunez K., et al. (2020) Diversity and global distribution of viruses of the western honey bee, *Apis mellifera, Insects*, 11: 239, https://doi.org/10.3390/insects11040239

30. Blanchard P., Ribiere M., Celle O., Lallemand P., et al. (2007) Evaluation of a real-time two step RT-PCR assay for quantitation of *Chronic bee paralysis virus* (CBPV) genome in experimentally-infected bee tissues and in life stages of a symptomatic colony, *J. Virol.* Methods, 141, 7-13.

31. Bowen-Walker P.L., Martin S.J., Gunn A. (1999) The transmission of deformed wing virus between honeybees (*Apis mellifera* L.) by the ectoparasitic mite *Varroa jacobsoni*, Oud. *J. Invertebr. Pathol.* 73, 101-106.

32. Brodschneider R., Crailsheim K. (2010) Nutrition and health in honey bees, *Apidologie* 41, 278-294.

33. Brodsgaarf C.J., Hansen H., Ritter W. (2000) Progress of *Paenibacillus larvae larvae* infection in individually inoculated honey bee larvae reared single in vitro, in micro colonies, or in full-size colonies, *J. Apic. Res.* 39, 19-27.

34. Browne K.A., Hassett J., Geary M., Moore E. et al. (2020) Investigation of free-living honey bee colonies in Ireland, *J. Apic. Res.*, https://doi.org/10.1080/00218839.2020.1837530

35. Buchler R. (1994) *Varroa* tolerance in honey bees - occurrence, characters and breeding, *Bee World* 49, 6-18.

36. Buchler R., Berg S., Le Conte Y.

(2010) Breeding for resistance to Varroa destructor in Europe, *Apidologie* 41, 393-408.

37. Budge G.E., Simcock N.K., Holder P.J., Shirley M.D.F. et al. (2020), Chronic bee paralysis as a serious emerging threat to honey bees, *Nature communications*, doi. org/10.1038/s41467-020-15919-0

38. Carreck N.L., Ball B.V., Martin S.J. (2010) The epidemiology of cloudy wing virus infections in honey bee colonies in the UK, *J. Apic. Res.* 49: 66-71.

39. Celle O., Blanchard P., Schurr F., Olivier V., et al. (2008) Detection of *Chronic bee paralysis virus* (CBPV) genome and RNA replication in various host: possible ways of spread, *Virus Res.* 133, 280-284.

40. Chen Y.P., Siede R. (2007) Honey bee viruses, Adv. *Virus Res.* 70, 33-80.

41. Chen Y., Zhao Y., Hammond J., Hsu H.T., Evans J., Feldlaufer M. (2004) Multiple virus infections in the honey bee and genome divergence of honey bee viruses, *J. Invertebr. Pathol.* 87, 84-93.

42. Cheshire F.R., Cheyne W.W. (1885) The pathogenic history and the history under cultivation of a new bacillus (*B. alvei*), the cause of a disease of the hive bee hitherto known as foul brood, J. Roy. *Microsc. Soc.* 5, 581-601.

43. Coffey M.F. (2007) Biotechnical methods in colony management, and the use of Apiguard® and Exomite™ Apis for the control of the varroa mite (Varroa destructor) in Irish honey bee (Apis mellifera) colonies, *J. Apic. Res.* 46(4): 213-219.

44. Coffey M.F. (2007) Parasites of the honeybee, DAFF, Dublin. 77 pp.

45. Coffey M.F., Breen J., Brown M.J.F., McMullan J.B. (2010) Brood-cell size has no influence on the population dynamics of *Varroa destructor* mites in the native European honey bee, *Apis mellifera mellifera*, *Apidologie* 41: 522-530.

46. Cox-Foster D.L., Conlan S., Holmes E.C., Palacios G., et al. (2007) A metagenomic survey of microbes in honey bee colony collapse disorder, *Science* 318: 283-287.

47. Crane E. (1999) The World History of beekeeping and Honey Hunting, Duckworth, London.

48. Crane E., Walker P. (1998) Irish beekeeping in the past, *Ulster Folklife* 44, 45-59.

49. Dade H.A. (1994) Anatomy and dissection of the honeybee IBRA, Cardiff

50. Dainat B., Ken T., Berthoud H., Neumann P. (2009) The ectoparasitic mite *Tropilaelaps mercedesae* (*Acari, Laelapidae*) as a vector of honeybee viruses, *Insect. Soc.* 56, 40-43.

51. Danforth B.N., Sipes S., Fang J., Brady S.G. (2006) The history of early bee diversification based on five genes plus morphology, *PNAS*, 103, 41, 15118-15123.

52. Danka R.G., Villa J.D. (1998) Evidence of autogrooming as a mechanism of honeybee resistance to tracheal mite infestation, *J. Apic. Res.* 37, 39-46.

53. Davis C.F. (2004) The honey bee inside out, *BeeCraft*

54. De Guzman L.I., Rinderer T.E., Delatte G.T., Stelzer J. A. et al. (2002) Resistance to *Acarapis woodi* by honey bees from Far-eastern

Russia., *Apidologie* 33, 411- 415.

55. Delfanado M.D., Baker E.W. (1961) Tropilaelaps, a new species of mite from the Philippines (Laelaptidae [s.lat]: Acarina), Fieldiana Zool 44, 53-56.

56. Delfinado-Baker M., Baker E.W. (1982) A new species of *Tropilaelaps* parasitic on honey bees, *Am. Bee J.* 122, 416-417.

57. de Miranda J.R., Genersch E. (2010) Deformed wing virus, *J. Invertebr. Pathol.* 103, S48-S61.

58. de Miranda J.R., Cordoni G., Budge G. (2010) The Acute bee paralysis virus-Kashmir bee virus-Israeli acute paralysis virus complex, *J. Invertbr. Pathol.* 103, S30-S45.

59. Digges J.G. (1904) The Irish Bee Guide. Eason and Son, Dublin. 220 pp.

60. Dobbs A. (1750) Bees and their method of gathering wax and honey, *Philosophical Transaction of the Royal Society*, 536-549.

61. Downey D.L., Winston M.L. (2001) Honey bee colony mortality and productivity with single and dual infestations of parasitic mite species, *Apidologie* 32, 567-575.

62. Doublet V., Labarussias M., De Miranda, J.R., Moritz R.F.A., Paxton J.R. (2014) Bees under stress: sublethal doses of a neonicotinoid pesticide and pathogen interact to elevate honey bee mortality across life cycle, *Environ Microbiol.*, doi:10.1111/1426-2920.12426

63. Eichwort G.C. (1988) The origin of mites associated with honey bees, in: Needham, G. R., Gage, R.E. Jr., Delfinado-Baker, M. and Bowman, C.E. (Eds.), Africanised honey bees and bee mites. Ellis Horwood, Chichester.

64. Eischen F.A. (1987) Overwintering performance of honeybee colonies heavily infested with *Acarapis woodi* (Rennie), *Apidologie* 18: 293-304.

65. Ellis A.M., Hayes G.W., Ellis J.D. (2009) The efficacy of dusting honey bee colonies with powdered sugar to reduce varroa mite populations, *J. Apic. Res.* 48, 72-76.

66. Ellis J. D. (2004) The ecology and control of small hive beetles (*Aethina tumida* Murray), PhD dissertation, Rhodes University, Grahamstown, South Africa.

67. Ellis J.D., Munn P.A. (2005) The worldwide health status of honey bees, *Bee World* 86, 88-101.

68. Ellis J. D., Delaplane K. S., Hood W. M. (2001) Small hive beetle (*Aethina tumida*) weight, gross biometry, and sex proportion at three locations in the Southeastern United States, *Am. Bee J.* 142 (7), 520-522.

69. Ellis J.D., Delaplane K.S., Hood W.M. (2001) Efficacy of a bottom screen device, Apistan, and Apilife VAR, in controlling *Varroa destructor*, *Am. Bee J.* 141, 813-816.

70. Ellis J.D., Richards C.S., Hepburn H. R., Elzen P. J. (2003) Oviposition by small hive beetles elicits hygienic responses from Cape honeybees, *Naturwissenschaften*, 90, 532-535.

71. Ellis M.D., Baxendale F. P. (1997) Toxicity of seven monoterpenoids to tracheal mites (*Acari: Tarsonemidae*) and their honey bee (Hymenoptera: Apidae) hosts when applied as fumigants, *J. Econ. Entomol.* 90, 1087-1091.

72. Elzen P. J., Baxter J. R.., Westervelt D., Randall C., et al. (1999) Field

control and biology studies of a new pest species, *Aethina tumida* Murray (Coleoptera, Nitidulidae) attacking European honey bees in the Western hemisphere, *Apidologie* 30, 361- 366.

73. Eyer M., Chen Y.P., Schafer M.O., Pettis J., et al. (2008) Small hive beetle, as a potential biological vector of honeybee viruses, *Apidologie*. doi: 10.1051/ apido:2008051.

74. NBU (2017) Foul brood diseases of honey bees, www.nationalbeeunit.com

75. NBU (2017) Adult bee diseases, Nosema, www.nationalbeeunit. com

76. NBU (2020) Managing Varroa, www.nationalbeeunit.com

77. Fleming G. (1871) Animal plagues: their history, nature and prevention, Chapman and hall, London.

78. Flores J.M., Ruiz J.A., Ruz J.M., Puerta F., et al. (1996) Effect of temperature and humidity of sealed brood on chalkbrood development under controlled conditions, *Apidologie* 27, 185-192.

79. Forsgren E., Fries I. (2010) Comparative virulence of *Nosema ceranae* and *Nosema apis* in individual European honey bees, *Vet. Parasitol*. 170, 212-217.

80. Forsgren E., de Miranda Joachim R., Isaksson Mats., Wei Shi., et al. (2009) *Deformed wing virus* associated with *Tropilaelaps mercedesae* infesting European honey bees (*Apis mellifera*), *Exp. Appl. Acarol*. 47: 87-97, doi10.1007/ s10493-008-9204-4

81. Frey E., Schnell H., Rosenkranz P. (2011) Invasion of Varroa destructor mites into mite-free honey bee colonies under controlled conditions of a military training area, *J. Apic. Res*. 50, 138-144.

82. Fries I. (1997) Protozoa, in: Morse, R.A., Flottum, K., (Eds.), Honey Bee Pests, Predators and Diseases, third ed. A.I. Root Company, Medina, Ohio, USA, pp.59-76.

83. Fries I. (2010) Nosema ceranae in European honey bees (*Apis mellifera*), *J. Invert. Pathol*. 103, s73-s79.

84. Fries I., Camazine S. (2001) Implications of horizontal and vertical pathogen transmission for honey bee entomology, *Apidologie* 32, 199-214.

85. Fries I., Imdorf A., Rosenkranz P. (2006) Survival of mite infested (*Varroa destructor*) honey bee (*Apis mellifera*) colonies in a Nordic climate, *Apidologie* 37, 564-570.

86. Fries I., Feng F., da Silva A., Slemenda S.B., et al. (1996) *Nosema ceranae* n. Sp. (Microspora, *Nosematidae*), morphological and molecular characterization of a microsporidian parasite of the Asian honey bee *Apis cerana* (Hymenoptera, Apidae), *Eur. J. Protistol*. 32, 356- 365.

87. Furgala B. (1962) The effect of the intensity of Nosema inoculum on queen supersedure in the honey bee, *Apis mellifera* Linnaeus, *J. Insect Pathol*. 4, 429-432.

88. Genersch E. (2011) American foulbrood in honeybees and its causative agent, *Paenibacillus larvae*, *J. Invertebr. Pathol*. 103: S10-S19.

89. Genersch E., Yue C., Fries I., de Miranda J.R. (2006) Detection of

*Deformed wing virus*, a honey bee viral pathogen, in bumble bees (*Bombus terrestris* and *Bombus pascuorum*) with wing deformities, *J. Invertebr. Pathol*. 91, 61-63.

90. Genersch E., Forsgren E., Pentikainen J., Ashiralieva A., et al. (2006) Reclassification of Paenibacillus larvae subsp. pulvifaciens and *Paenibacillus larvae* subsp. larvae as *Paenibacillus larvae* without subspecies differentiation, Int. *J. Syst. Evol. Microbiol*. 56, 501-511.

91. Gill R.J., Ramos-Rodriguez O., Raine N.E. (2012) Combined pesticide exposure severely affects individual - and colony-level traits in bees, *Nature* 491:105-108, doi. org/10.1038/nature11585

92. Gilliam M. (1997) Identification and roles of non- pathogenic microflora associated with honey bees, *FEMS Microbiol*. Lett. 155, 1-10.

93. Glinski Z. (1982) Studies on pathogenicity of Ascosphaera apis for larvae of the honeybee *Apis mellifera* L. Part 11. Relationships between biochemical types and virulence of A. apis, Annales Universitatis Mariae Curie Sklodowska Sectio DD. *Med. Vet*. 37(8), 69-77.

94. Goodwin M., Van Eaton C. (1999) Elimination of American foulbrood without the use of drugs: A practical manual for beekeepers, National Beekeepers' Association of New Zealand, Inc 78 pp.

95. Greatti M., Milani N., Nazzi F. (1992) Reinfestation of an acaricide-treated apiary by *Varroa jacobsoni* Oud., Exp. Appl. Acarol. 16, 279-286.

96. Gregory P. (2011) All about Nosema, *Bee Craft* 93, 9, 33-36.

97. Han F., Wallberg A. Weber M.T. (2012) From where did the Western honeybee (*Apis mellifera*) originate? *Ecology and evolution,* 2(8): 1949-1957.

98. Hansen H., Brodsgaard C.J. (1999) American foulbrood: a review of its biology, diagnosis and control, *Bee World* 80, 5-23.

99. Harbo J.R, Harris J.W. (2004) Effect of screened floors on populations of honey bees and parasitic mites (*Varroa destructor*), *J. Apic. Res*. 43, 114-117.

100. Harrison J.F., Camazine S., Marden J.H. et al. (2001) Mite not make it home: Tracheal mites reduce the safety margin for oxygen delivery of flying honeybees, *J. Exp. Biol*. 204: 805-814.

101. Hasset J., Brown K.A., McCormack G.P., Moore E., et al. (2018) A significant pure population of the dark European honey bee (*Apis mellifera mellifera*) remains in Ireland, *J. Apic. Res*. 57(3): 337-350, https.//doi.org/10.1080/00218839.20 18.1433949

102. Heyndrickx M., Vandemeulebroecke K., Hoste B., Janssen P., et al. (1996) Reclassification of *Paenibacillus* (formerly *Bacillus*) *pulvifaciens* (Nakamura 1984) Ash et al. 1994, a later subjective synonym of larvae, with amended description of *P. larvae* as *P. larvae* subsp. larvae and P. larvae subsp. Pulvifaciens, International *Journal of Systematic Bacteriology* 46: 270-279.

103. Higes M., Martin-Hernandez R., Botias C., Bailon E.G., et al. (2008) How natural infection by *Nosema ceranae* causes honeybee colony

collapse, *Environ. Microbiol*. 10, 2659-2669.

104. Highfield A.C., El Nagar A., Mackinder L.C.M., Noel L.M.L.J., et al. (2009) Deformed wing virus implicated in overwintering honey bee colony losses, *App. Environ. Microbiol*. 75: 7212-7220.doi: 10.1128/AEM.02227-09.

105. Hooper T. (1997) A guide to bees and honey, Marston House, Somerset, UK. 271 pp. ISBN 1-899296-04-2.

106. Ibrahim A., Reuter G.S., Spivak M. (2007) Field trial of honey bee colonies bred for mechanisms of resistance against Varroa destructor, *Apidologie* 38 (1), 67-76.

107. Jandricic, S.E., Otis, G.W. (2003) The potential for using male selection in breeding honey bees resistant to *Varroa destructor. Bee World* **84**(4), 155-164.

108. Johansson T.S.K., Johansson M.P. (1978) Some important operations in bee management, International Bee Research Association, Bucks, England. pp 70-88.

109. Kaskoniene V., Maruska A., Kornysova O., Charczun N. et al. (2009) Quantitative and qualitative determination of phenolic compounds in honey, *Chemine Technologija*, 3, 74-80, doi. org/10.1016/S0308-8146(96)00313-5

110. Kavanagh S., Gunnoo J., Passos T.M., Stout J.C. et al. (2019) Physicochemical properties and phenolic content of honey from different floral origins and from rural versus urban landscapes. *Food Chemistry* 272: 66-75. https://www. sciencedirect.com/science/article/pii/ S0308814618314365?via%3

Dihub

111. Kirk W. (1994) A colour guide to pollen loads of the honey bee, International Bee Research Association, Cardiff 54 pp.

112. Klee J., Besana A.M., Genersch E., Gisder S., et al. (2007) Widespread dispersal of the microsporidian Nosema ceranae, an emergent pathogen of the western honey bee, *Apis mellifera, J. Inverteb. Pathol*. 96, 1-10.

113. Kojima Y., Toki T., Morimoto T., Yoshiyama M., et al. (2011) Infestation of Japanese native honey bees by tracheal mite and virus from non-native European honey bees in Japan, *Microb. Ecol.* 62: 895-906 DOI 10.1007/ s00248-011-9947-z

114. Komeili A.B., Ambrose J.T. (1991) Electron microscope studies of the tracheae and flight muscles of noninfested, *Acarapis woodi* infested and crawling honey bees *Apis mellifera), Am Bee J.* 131: 253-257.

115. Korpela S., Aarhus A., Fries I., Hansen H. (1993) *Varroa jacobsoni* Oud. In cold climates: population growth, winter mortality and influence on survival of honey bee colonies, *J. Apic. Res.* 31, 157-164.

116. Kraus F.B., Neuman P., Moritz R.F.A. (2005) Genetic variance of mating frequency in the honeybee (*Apis mellifera L.*), *Insectes Soc.* 52, 1-5, https://doi.org/10.1007/s000040-004-0766-9

116a. Kronenberg A.B., Heller C. (1982) Colonial thermoregulation in honey bees (*Apis mellifera*), *J. Comp. Physio*. 148, 65-76.

117. Kulincevic J.M., Rothenbuhler W.C. (1975) Selection for resistance

and susceptibility to hairless-black syndrome in the honey bee, *J. Invertebr. Pathol.* 25, 289-295.

118. Le Conte Y., Arnold G. (1988) *Etude du Thermopreferendum de Varroa Jacobsoni* (Oud). *Apidologie*, 19, 155-164.

119. Le Conte Y., de Vaublanc G., Crauser D., Jeanne F., et al. (2007) Honey bee colonies that have survived Varroa destructor, *Apidologie* 38, 566-572.

120. López J.H., Krainer S., Engert A., Schuehly W. et al. (2017) Sublethal pesticide doses negatively affect survival and the cellular responses in American foulbrood-infected honeybee larvae, *Sci. Rep.* 7:40853, doi: 10.1038/srep40853

121. Lundie A. E. (1940) The small hive beetle, Aethina tumida, South Africa Department of Agriculture & Forestry Science Bulletin no. 220. 30pp.

122. Maassen A. (1913) Weitere Mitteilungen uber der seuchenhaften Brutkrankheiten der Bienen (Further communication on the epidemic brood disease of bees), Mitteilungen aus der *Kaiserlichen Biologischen Anstalt fur Land und Forstwirtscshaft* 14, 48-58.

123. Martin S.J. (2001) The role of Varroa and viral pathogens in the collapse of honeybee colonies: A modelling approach, J. Appl. Ecol. 38, 1082-1093.

124. Mattila H.A., Seeley T.D. (2014) Extreme polyandry improves a honey bee colony's ability to track dynamic foraging opportunities via greater activity of inspecting bees, *Apidologie* 45, 347-363

125. Maurizio A. (1934) Uber die Kaltbrut (Pericystis-Mykose) der Bienen, *Archiv Bienenkunde* 15, 165-193.

126. McMullan J. (2000) Weighing beehives: a simple lever method, *Bee World* 81, 11-19.

127. McMullan J.B. (2011) Migration of tracheal mites (*Acarapis woodi*) to old honey bees, *Apidologie* 42, 577-578.

128. McMullan, J. (2012) Having healthy honeybees, an integrated approach. FIBKA, Dublin, ISBN 978-0-9571355-0-5.

129. McMullan, J. (2018). Adaptation in honeybee (*Apis mellifera*) colonies exhibiting tolerance to Varroa destructor in Ireland. *Bee World*, 95(2), 39–43, doi:10.1080/000577 2X.2018.1431000

130. McMullan J.B., Brown M.J.F. (2005) Brood pupation temperature affects the susceptibility of honeybees (*Apis mellifera*) to infestation by tracheal mites (*Acarapis woodi*), *Apidologie* 36, 97-105.

131. McMullan J.B., Brown M.J.F. (2006) The role of autogrooming in the differential susceptibility to tracheal mite (Acarapis woodi) infestation of honeybees (Apis mellifera) held at both normal and reduced temperatures duration pupation. *Apidologie* 37, 471-479.

132. McMullan J.B., Brown M.J.F. (2006) Brood-cell size does not influence the susceptibility of honeybees (*Apis mellifera*) to infestation by tracheal mites (*Acarapis woodi*), *Exp. Appl. Acarol.* 39, 273-280.

133. McMullan J.B., Brown M.J.F. (2006) The influence of small-cell brood combs on the morphometry of honeybees (Apis mellifera L.),

*Apidologie* 37, 665-672.

134. McMullan J.B., Brown M.J. F. (2009) A qualitative model of mortality in honey bee (*Apis mellifera*) colonies infested with tracheal mites (*Acarapis woodi*), *Exp. Appl. Acarol.* 47, 225-234.

135. McMullan J.B., d'Ettorre P., Brown M.J.F. (2010) Chemical cues in the hostseeking behaviour of tracheal mites (*Acarapis woodi*) in honey bees (Apis mellifera mellifera), *Apidologie* 41, 568-578.

136. McMullan J. (2006) A simple approach to weighing beehives, Nutshell No. 85. Northern Bee Books, Hebden Bridge, UK. 6pp.

137. McMullan J.B. (2008) Tracheal mite (acarine) infestation in honeybees - some recent research, The Central Association of Beekeepers, Ilford, UK.

138. McMullan J. (2008) Acarine in honey bees, *Bee Craft* 90, 6, 13-16.

139. McMullan J. (2008) Feeding honey bee colonies, *Bee Craft* 90, 9, 21-23.

140. McMullan J. (2010) Mortality in tracheal-mite-infested colonies and the role of thermoregulation, *Am. Bee J.* 150, 165-169.

141. McMullan J. (2011) The beginner beekeeper, Bee Craft 93, 9, 30-31.

141a. McMullan J. (2019) A beekeeping regime that facilitates Varroa tolerance in honeybees, NIHBS.

142. McMullan J. (2013) The impact of research on beekeeping practices over the past 100 years, Proceedings 2013 Congress, *Apimondia*, Kiev, Ukraine

143. McMullan, J. (2020). "Is COVID-19 still the killer it was?", *The Sunday Times* 13th September 2020.

144. Meana A., Martin-Hernandez R., Higes M. (2010) The reliability of spore counts to diagnose Nosema ceranae infections in honey bees, *J. Apic. Res.* 49(2), 212-214.

144a. Meixner M.D., Buchler R., Costa C., Francis R.M. et al. (2014) Honey bee genotypes and the environment, *J. Apic. Res.*.53, 183-187. doi 10.3896/IBRA.1.53.2.01

145. Michener C.D. (1974) The social behaviour of the bees: a comparative study, Cambridge, Mass., Harvard Univ. Press.

146. Morse R.A. (1994) Complete Guide to Beekeeping, The Countryman Press, Vermont, US. 203pp.

147. Murray A. (1867) List of Coleoptera received from Old Calabar, *Ann. Magazine Nat. Hist.*, London, 19, 167-179.

148. National Biodiversity Data Centre. (2016) All- Ireland Pollinator Plan 2015-2020, www.biodiversity.ireland.ie

149. Native Irish Honey Bee Society. www.nihbs.org

150. Naug D. (2009) Nutritional stress due to habitat loss may explain recent honeybee colony collapses, *Biol. Conserv.* 142, 2369-2372.

151. Neumann P., Pirk C.W.W., Hepburn H.R., Solbrig A.J. et al. (2001) Social encapsulation of beetle parasites by Cape honeybee colonies (*Apis mellifera capensis* Esch,), *Naturwissenschaften* 88, 214-216.

152. Nielsen S.L., Nicolaisen M., Kryger P. (2008) Incidence of acute bee paralysis virus, black queen cell virus, chronic bee paralysis virus, deformed wing virus, Kashmir bee virus and sacbrood virus in honey bees (Apis mellifera) in Denmark, *Apidologie* 39, 310-314.

153. Nordstrom S., Fries I., Aarhus A., Hansen H., Korpela S.(1999) Virus

infections in Nordic honey bee colonies with no, low or severe Varroa jacobsoni infestations, *Apidologie* 30, 475-484. doi: 10.1051

154. Oldroyd B. P. (1999) Coevolution while you wait: *Varroa jacobsoni* a new parasite of western honeybees, *Trends Ecol. Evol.* 14: 312-315.

155. OIE Terrestrial Manual (2008) Chapters 2.2.2., 2.2.3.

156. Olivier V., Blanchard P., Chaouch S., Lallemand P., et al. (2008) Molecular characterization and phylogenetic analysis of Chronic bee paralysis virus, a honey bee virus, *Virus Res.* 132, 59-68.

157. Otis G.W., Scott-Dupree C.D. (1992) Effects of *Acarapis woodi* on overwintered colonies of honey bees (*Hymenoptera: Apidae*) in New York, *J. Econ.* Entomol 85: 40-46

158. Owens C.D. (1971) The thermology of wintering honey bee colonies, *Tech. Bull. U.S. Dep. Agric.* 1429, 1-32.

159. Paxton R.J. (2010) "Does infection by Nosema ceranae cause 'Colony Collapse Disorder in honey bees (*Apis mellifera*)? *J. Apic. Res.* 49, 80-84.

160. Paxton R.J., Klee J., Korpela S., Fries I. (2007) Nosema ceranae has infected *Apis mellifera* in Europe since at least 1998 and may be more virulent than *Nosema apis*, *Apidologie* 38, 558-565.

161. Pentos K., Luczycka D., Oszmianski J., Lachowicz S. et al. (2020) Polish honey as a source of antioxidants – a comparison with Manuka honey, *J. Apic. Res.* 59, 939-945, https://doi. org/10.1080/00218839.2020.1723837

162. Pettis J.S., Pankiw T. (1998) Grooming behavior by *Apis mellifera* L. in the presence of *Acarapis woodi* (Rennie) (Acari: Tarsonemidae), *Apidologie* 29, 241-253.

163. Planck M. (1950) scientific autobiography, and other papers, Williams and Norgate.

164. Ramsey S.D., Ochoa R., Bauchan G., Gulbronson C., et al. (2019) *Varroa destructor* feeds primarily on honey bee fat body tissue and not hemolymph, *PNAS* https://doi. org/10.1073/pnas.1818371116

165. Rennie, J. (1921). Isle of Wight disease in hive bees – acarine disease: the organism associated with the disease – *Tarsonemus woodi. Transcripts of the Royal Society of Edinburgh* 52: 768-779.

166. Ribiere M., Ball B., Aubert M. (2008) Natural history and geographical distribution of honey bee viruses, in: Aubert M., Ball B., Fries I., Moritz R. F., Milani N., Bernardinelli I . (Eds.), *Virology and the honey bee*, European Commission, Brussels.

167. Ribiere M., Olivier V., Blanchard P. (2010) Chronic bee paralysis: A disease and a virus like no other? *J. Invertebr. Pathol.* 103, s120-s131.

168. Rinderer T. E., Rothenbuhler W. C. (1975) Responses of three genetically different stocks of the honeybee to a virus from bees with hairless-black syndrome, *J. Invertebr. Pathol.* 25, 297-300.

169. Rinderer T.E., Harris J.W., Hunt G. J., de Guzman L.I. (2010) Breeding for resistance to Varroa destructor in North America, *Apidologie* 41, 409-424.

170. Roetschi A., Berthoud H., Kuhn R., Imdorf A. (2008) Infection rate

based on quantitative real-time PCR of *Melissococcus plutonius*, the causal agent of European foulbrood, in honeybee colonies before and after apiary sanitation, *Apidologie* 39, 362-371.

171. Rosenkranz P., Aumeier P., Ziegelmann B. (2010) Biology and control of *Varroa destructor*, *J. Invertebr. Pathol.* 103, S96- S119.

172. Royal Botanic Gardens. Kew (2020) State of world's plants and fungi. doi.org//10,34885/172.

173. Runckel C., Flenniken M.L., Engel J.C., Ruby J.G. et al. (2011) Temporal analysis of the honey bee microbiome reveals four novel viruses and seasonal prevalence of known viruses, *nosema*, and *crithidia*, *PloS ONE* 6 (6): e20656. doi: 10.1371/journal.pone.0020656

174. Rundlof M., Anderson G.K.S., Bommarco R., Fries I., et al., (2015) Seed coating with a neonicotinoid insecticide negatively affects wild bees, *Nature* 521: 77-80. doi: 10.1038/nature11585

175. Sakamoto Y., Maeda T., Yoshiyama M., Pettis J.S. (2017) Differential susceptibility to the tracheal mite *Acarapis woodi* between *Apis cerana* and *Apis mellifera*, *Apidologie* 48, 150-158.

176. Sammataro D., Cobey S., Smith B.H., Needham G. R. (1994) Controlling Tracheal Mites (Acari: Tarsonemidae) in Honey Bees (Hymenoptera: Apidae) with Vegetable Oil, *J. Econ. Entom.* 87(4): 910-916 (1994).

177. Sammataro D., Untalan P., Guerrero F., Finley J. (2005) The resistance of Varroa mites (Acari: Varroidae) to acaricides and the presence of esterase, *Int. J. Acarol.* 31, 67-74.

178. Samuelson A.E., Gill R.J., Leadbetter E. (2020) Urbanisation is associated with reduced Nosema sp. Infection, higher colony strength and higher richness of foraged pollen in honeybees, *Apidologie* 51, 746-762. doi: 10.1007/sl3592-020-00758-1

179. Schafer M.O., Pettis J.S., Ritter W., Neumann P. (2010) Simple small hive beetle diagnosis, *Am. Bee J.* 150, 371-372.

180. Schmid-Hempel P. (1998) Parasites in social insects, Princetown University Press, New Jersey.

181. Seeley T.D. (1985) Honeybee Ecology, Princeton University Press, Princeton.

182. Seeley T.D. (1989) The honeybee colony as a superorganism, *American naturalist* 77: 546-553.

183. Seeley T. (2007) Honey bees of the Arnot Forest: a population of feral colonies persisting with Varroa destructor in the north eastern United States, *Apidologie* 38, 19-29.

184. Seeley T.D., Griffin S.R. (2011) Small-cell comb does not control Varroa mites in the colonies of honeybees of European origin, *Apidologie* 42, 526-532.

185. Seeley T.D., Morse R.A. (1976) The nest of the honey bee (*Apis mellifera* L.), *Insectes Sociaux* 23, 495-512.

186. Seeley T.D., Visscher P.K. (1985) Survival of honeybees in cold climates: the critical timing of colony growth and reproduction, *Ecol. Entomol.* 10,

187. Seeley, T.D., Tarpy, D.R., Griffin, S.R., Carcione, A., Delaney, D.A. (2015) A survivor population of wild colonies of European

honeybees in northeastern United States: investigating its genetic structure. *Apidologie* **46** (5), 559-67881-88.

188. Shimanuki, H., Knox, D.A. (2000) Diagnosis of honey bee diseases. U.S. Dept. Agric Handbook No. 690.

189. Simone-Finstrom M., Spivak M. (2010) Propolis and bee health: the natural history and significance of resin use by honey bees, *Apidologie* 41, 295-311.

190. Sina M., Alastair G., Farmer M., Andersen R., et al (2005) The New Higher level classification of Eukaryotes with emphasis on the taxomony of Protists, *J. Eukar. Microbiol.* 52, 399-451.

191. Snodgrass R. E. (1956) Anatomy of the honey bee, Cornell University Press Ltd., London

192. Spiltoir C.F. (1955) Life cycle of Ascosphaera apis, *Am. J. Bot.* 42, 501-518.

193. Spiltoir C.F., Olive L.S. (1955) A reclassification of the genus Pericystis betts, *Mycologia* 47, 238-244.

194. Spivak M., Gilliam M. (1993) Facultative expression of hygienic behaviour of honeybees in relation to disease resistance, *J. Apic. Res.* 32, 147-157.

195. Stanley D., Gunning D., and Stout J. (2013) Pollinators and pollination of oilseed rape crops (*Brassica napus* L.) in Ireland: ecological and economic incentives for pollinator conservation. *J. Insect Conser,* 17: 1181-1189.

196. Stephens W. (1733) Instructions for managing bees, Dublin Society, Dublin.

196a. Storch H. (1985) At the hive entrance. *European Apicultural Editions*

197. Suazo A., Torto B., Teal P.E., Tumlinson J.H. (2003) Response of the small hive beetle (*Aethina tumida*) to honey bee (*Apis mellifera*) and bee- hive produced volatiles, *Apidologie* 34, 525-533.

198. Sumpter D.J.T., Martin S.J. (2004) The dynamics of virus epidemics in Varroa- infested honey bee colonies, *J. Anim. Ecol.* 73, 51-63.

199. Tarpy D.R., Seeley T.D. (2006) Lower disease infections in honeybee (*Apis mellifera*) colonies headed by polyandrous vs monandrous queens, *Naturwissenschaften* 93, 195-199.

200. Tautz, J., Maier, S., Groh, C., Rossler, W. et al. (2003). Behavioral performance in adult honey bees is influenced by the temperature experienced during their pupal development. *PNAS* 100, 343-7347.

201. Tentcheva D., Gauthier L., Zappulla N., Dainat B., et al. (2004) Prevalence and seasonal variations of six bee viruses in *Apis mellifera* L. and *Varroa destructor* mite populations in France, *Appl. Environ. Microbiol.* 70, 7185-7191.

202. Theantana T., Chantawannakul P. (2008) Protease and ß-N acetylglucosaminidase of honeybee chalkbrood pathogen *Ascosphaera apis*, *J. Apic. Res.* 47, 68-76.

203. Thompson H.M., Waite R.J., Wilkins S., Brown M.A., et al. (2006) Effects of shook swarm and supplementary feeding on oxytetracycline levels in honey extracted from treated colonies, *Apidologie* 37, 51-57.

204. Tomkies V., Flint J., Johnson G., Waite R., et al. (2009) Development and validation of a novel field test kit for European foulbrood,

*Apidologie* 40, 63-72.

205. Traynor K.S., Rennich K., Forsgren E., Rose R. (2016) Multiyear survey targeting disease incidence in US honey bees, *Apidologie* 47, 325-347, doi: 10.1007/s13592-016-0431-0

206. Truper H.G., de Clari L. (1998) Taxonomic note: erratum and correction of further specific epithets formed as substantives (nouns) in apposition, *Int. J. Syst. Bacteriol.* 48. 615.

207. UK National Ecosystem Assessment: Northern Ireland Summary (2011) http:/www.nienvironmentlink. org/cmsfiles/policy/hub/files/ documentation/Eco/Northern-Ireland-NEA

208. van Engelsdorf D., Evans J.D., Saegerman C., Mullin C., et al. (2009) Colony collapse disorder: a descriptive study, *PloS ONE* 4 (8), e6481. doi:10. 1371/journal. pone.0006481

209. Vidau C., Diogon M., Aufauvre J., Fontbonne R., et al. (2011) Exposure to sublethal doses of fipronil and thiacloprid highly increases mortality of honeybees previously infected by *Nosema ceranae*, PLoS ONE 6, e21550.

210. Wallberg A., Han F., Wellhagen G., Dahle B. (2014) A worldwide survey of genome sequence variation provides insight into the evolutionary history of the honeybee *Apis mellifera*, *Nature Genetics*, doi: 10.1038/NG.3077

211. Waring A., Waring C. (2011) Bee Manual. Haynes, Somerset. 169 pp. ISBN 978 0 85733 057 4

212. Watson J. (1981) Bee-Keeping in Ireland, Glendale Press, Dublin. 293 pp.

213. White G. F. (1906) The bacteria of the apiary with special reference to bee disease, US Dept. *Agric., B. Entomol.*, T.S. No.14.

214. White G. F (1907) The cause of American foulbrood, US Dept. Agric., *B. Entomol.,* Circular 94.

215. White G. F (1912) The cause of European foulbrood, US Dept. Agric., *B. Entomol.*, Circular 157.

216. White G.F. (1917) Sacbrood, B. *US Dept. Agric.*, No. 431.

217. Williams G.R., Shutler D., Little C.M., Burgher- MacLellan K. L., et al. (2010) The microsporidian Nosema ceranae, the antiobiotic Fumagilin-B,and western bee (Apis mellifera) colony strength, *Apidologie* 42, 15-22.

218. Winston, M.L. (1987) The biology of the honeybee. Havard University Press, Cambridge, Massachusetts. 267pp.

219. Woodcock B.A., Isaac N.J.B., Bullock J.M., Roy D.B., et al. (2016) Impact of neonicotinoid use on long-term population changes in wild bees in England, *Nature Comm.* doi:10.1038/ncomms12459

220. Woyke J (1984) Survival and prophylactic control of *Tropilaelaps clareae* infesting *Apis mellifera* colonies in Afghanistan, *Apidologie* 15: 421-434.

221. Woyke J. (1987) Length of successive stages in the development of the mite *Tropilaelaps clareae* in relation to honeybee brood age, *J. Apic.Res.* 26110-114.

222. Yang X., Cox-Foster D. (2007) Effects of parasitization by Varroa destructor on survivorship and physiological traits of Apis mellifera in correlation with viral incidence and microbial challenge,

*Parasitology* 13, 405-412.

223. Yue C., Schroder M., Gisder S., Genersch E. (2007) Vertical transmission routes for deformed wing virus of honeybees (*Apis mellifera*), *J. Gen. Virol.* 88, 2329-2336.

224. Yue D., Nordhoff M., Wieler L.H., Genersch E. (2008) Fluorescence in situhybridization (FISH) analysis of the interactions between honeybee larvae and Paenibacillus larvae, the causative agent of American foulbrood of honeybees (*Apis mellifera*), *Environ. Microbiol.* 10, 1612-1620.

225. Zander E. (1909) Tierische Parasiten als Krankheitserreger bei der Biene, *Leipziger Bienenztg* 24, 147-150, 164-166.

# Index